4HourUX

An End-to-End Framework for Designing User Experiences

Sarah Deane

Mention of any specific companies, organizations, or authorities in this book does not imply endorsement by the author or publisher, nor does the mention of any specific companies, organizations, or authorities imply that they endorse this book, its author, or the publisher.

All facts referenced in this book were correct at the time it was written.

ISBN: 978-0-9904140-0-1

DEDICATION

To my Mum, Dad, Brother and Sister for always being supportive of all my many ideas, never constraining my thoughts, and for inspiring me through their amazingness.

To K, for always being there for me, supporting me and believing in me. This wouldn't have been possible without you.

ACKNOWLEDGMENTS

A huge thanks to Allison Kluger and Mark Shaw, for their help and guidance. You are inspirational people that have great energy, minds and hearts. How you do it all I have no idea, but you do, and I cannot thank you enough!

***

A huge thanks to my close friends for all of their support and help in creating this book – I couldn't have done this without you (you know who you are).

CONTENTS

Preface

A long time ago, while still in my previous career of A.I and Image Processing, I was about to embark on a new career in User Experience (UX). I started reading up on the discipline to supplement what I already knew on the subject from prior studies and professional experience.

There was a wealth of information on design and UX and it all seemed rather overwhelming. Not to mention, the world was rapidly changing bringing with it huge paradigm shifts in how people interacted with technology. I wrote out guidelines to follow that I thought would make sense in this world in hopes of bringing UX to an organization in my new role. Now many years later, I have refined these guidelines with learnings and other nuggets of valuable information gathered along the way, which I have proven can work and which shows how you can go from limited or no UX consideration, to having a team, an organization and a practice. Seeing the UX discipline erupt within companies who are now considering UX as a differentiator, and with UX as a job role in high demand and growing rapidly, it was time to share these learnings with anyone who may need it.

This guide will take you on a journey from start to finish to show you how to think about UX. It glues together all the components so that you are equipped with the fundamentals you will need to create user experiences that work. It is meant for those that are new to UX or those who find themselves in a UX job and do not know where to start. This book is also for those in disciplines such as marketing and communications who need to understand this new world where UX is key, and those who are responsible for institutionalizing the UX discipline in teams, organizations or companies including employee experience. Whether you are a CXO, a hotel owner, a product manager, a marketer, or a UX professional, it will provide you with the knowledge of how to understand your product's experience and how to create a great one from start to finish.

4 Hour UX aims to provide a quick-to-read, end-to-end structured overview that will equip you with a good amount of practical knowledge that you can put to use immediately. In fact, in true UX fashion I had the book pre-read by a representative group of my target audience, as well as throughout development, for their feedback. Two of them told me they read it on the 4 hour flight from San Francisco to Houston – hence the name.

I hope you enjoy the read as much as I enjoyed creating it.

1

ASPECTS OF UX TODAY

There are many books, papers, articles, and even companies all writing about experience design. However, sometimes it can be difficult to know how to apply these theories, or how to put them into practice without seeing how all the dots are connected throughout the whole lifecycle of creating experiences.

I have created a framework to follow to take you step-by-step through the thought process from strategy, user testing to delivery. This UX primer allows you to see how all of the components fit together, providing a quick and practical overview, after which all the other great resources available can be better understood and utilized to focus on certain aspects of UX if needed.

Many companies, given the changes to the world today, are starting to focus more and more on the experience they give their users, but given normal company constraints such as resources and funding, much of the work is given to current employees who may not come from an appropriate discipline. Also, while an organization is starting in the UX space, leaders and managers may not have yet grasped what UX means or what needs to be done. The way to demonstrate and convince leadership new to the space is to show proven results, and to show them that the work required to have a UX focus is achievable without large overhead. Therefore starting out with some basic work in the space, and showing that it works and is within budget, is a first step. As confidence and belief grows throughout the organization, more work can be done and so on and so forth. This brings light to a core fact that a part of the role starting out is to influence the organization to internalize experience, what it means, and how to accomplish it.

This book marries many great principals of this space with a practical framework to follow. It is based on the reading of books and academic research in the area, listening to and talking with leaders in the experience domain from successful companies, as well as the knowledge gained via implementing such practices in a large corporation and small businesses myself.

It outlines the mindset, thought process and steps which will enable you to actually start doing some work in the space after reading it. It will equip you with a good level of practical knowledge as well as key theory's from the domain so that you can present and advocate with confidence to teams, management and executives (a core skill if you are to successfully institutionalize this practice into your organization).

You will see there is focus on enterprise user experience examples, or 'internal experiences' to an organization - the reason being that this is a great example of an organization new to the space with severe budget constraints. Also, it gives the opportunity to highlight some considerations for you to think about if you are creating experiences for business users as opposed to consumers. People that are new to the UX discipline or are in organizations with similar budget constraints can easily apply the principals into the design of anything.

THE CHANGING UX LANDSCAPE

For many years visual design ruled much of the experience landscape. While the UI and visual design still plays an extremely large role in experience today, there are several factors to consider including context, content, devices, the environment the user is in, the user themselves and how all the pieces are connected. **Experiences today have to be intuitively intelligent, adapting to the users situation, making them smarter, faster, more engaged and more efficient.** The hyper-connected, online world we now live in, and with the collision of the digital and physical worlds a reality – experience design is the glue. Experience designers today have to understand how everything plays together to create the best experience for their users from start to finish.

FOR THOSE WORKING ON ENTERPRISE USER EXPERIENCE SPECIFICALLY

Consumer experience is a well explored space and companies have for years spent millions of dollars into learning their consumer and how to get them to purchase their product. I'll never forget a conversation I had with a friend once that I think summarizes just how well some companies have become at what can almost be defined as 'consumer manipulation'. My friend was not that impressed with a large technology company's latest release. For years, this company had been a favorite in design and innovation, but to many (including my friend) their latest release was almost expected, nothing that great and some elements were almost hard-to-work-out. From his perspective, he knew it was not particularly well-designed, but he made a very interesting comment – 'that the company had their consumers so well trained that most people that couldn't use it didn't blame the design, instead they thought it was their fault since the company's products were always well designed and usable'. This demonstrates just how much some consumer product division's focus on their experience, their brand and how they understand their base. Historically there hasn't been as much investment in employee experience. However, recently there has been more and more focus in this space, as well as a growing appreciation of the link between good employee experiences leading to a great customer experience. There has been a huge rise in jobs specific to employee experience or engagement, so I have tried to include some nuggets specific to this space too.

There are many reasons why there has been an increasing boom in concentration on employee experience, however one of the main reasons that has been attributed is historical; how corporations have developed over time has led to this new focus.

Currently, there is a crisis in many organizations; they have a vast amount of detached, ready-to-exit employees. This has been described as a 'psychological recession' and a key reason why companies have started to focus more in the space. For several years companies have been trying to do more with less in order to drive greater profits and this has left an impact in the internal functions (functions that keep the business going i.e. Finance, IT, Facilities etc.) where they are a cost to the business as opposed to profit making. It has been stated that there are many big companies who have been around for a while that need the emphasis on employee experience, but also that those that become public will be subjected to similar corporate mechanistic views, once they start having to account every penny to stakeholders. There are many case studies of companies that have gone through a period of acquisitions, cost cutting, reorganizing business and new CEO's, which are typical traits of companies now starting to focus more on employee experience. These events lead to instability, which in turn lead to a shattered employee experience if not managed correctly.

One could say that the principals of the 20th century have been somewhat turned on their head, and that as we move into the next century, multiple changes are occurring in experience design. The previous century saw bureaucracy progress, leading to the dehumanization of the workplace. This led organizations to favor rationality and process to increase efficiency at the cost of their employees. A world of output optimization ensued; even pay for performance was introduced to many companies to get employees focused on productivity. Now, the humanization between clients, end users and employees has become a must for successful organizations. This 20th century thinking was focused on rationalizing organizations, without consideration to the emotional needs of the employees. Emotions are fighting back big time and those companies that focus on their employee needs have been seen to experience better brand perception as well as better a consumer experience. **It has now become increasingly more evident that there is a strong relationship between a positive employee experience and a positive consumer experience.**

In the 20th century, design was more a process to create tangible outputs for an organization, now, design is evolving to include things such as services and experiences. As organizations seek ways to increase their competitive advantage, the design of experience has become important in the consumer world, and now, just as important in the employee world. In fact, companies with engaged employees tend to exceed their industry's annual average growth in revenue by at least one percentage point, and in the times of highly competitive markets and recession, this is a huge factor in the emphasis on employee experience.

The truth is it can be extremely difficult to maneuver one's self into the internal world or into a team just starting out in this space, where an organization may not shell out dollars in studies of learning, or allow for hours of observations. Companies that are just starting to focus on user experience are likely to have extremely lean teams and not a great deal of budget to invest in their experience initiatives. Also, some companies are not willing to invest lots of money into user experience once maximizing shareholder return and corporate politics come into play. However, we are in a period of change, where companies are starting to focus on the space more and so in this time it is important to focus on practical ways to lead to an improved experience. Sadly the other factor is that some organizations simply do not know how to do it right, they think that having a period of 'brand-washing' is enough to bring their employees into a satisfied and motivated state or to create a product experience for their consumers. **When getting started, it is key to start out lean, prove the process works by demonstrating tangible value, and then start expanding throughout the organization.** This journey, is what some call 'institutionalizing' experience design.

So there you have it, employee experience is a fantastic example of usually lean teams, low cost environments and high pressure, creating a great basis to learn fundamentals that you can apply to the design of anything when operating in spaces with similar characteristics. Another important observation is that in fact, the employee experience and consumer experience worlds are blending. Employees are starting to expect the consumer experience they live with every day, to be matched internally too. As these worlds collide, it makes even more sense that the learning's can be shared. Luckily, there are some fundamentals that an organization can use, that are actually achievable on a limited budget, and will allow for a 'fat trimmed' experience design process.

KEY OBSERVED DIFFERENCES IN THE EXTERNAL AND INTERNAL WORLDS

While some companies do spend a substantial amount of time and money on creating experiences for a content and satisfied user base, many do not have the capability in knowledge, nor the funds to do so. There are key differences between them that pertain to companies in this situation, or companies who are just starting to focus on the experience.

A core take-away is that the employee doesn't want to buy a solution, more often than not, they want to do their job, be productive, and be enabled to have a fun and enriching work environment. This can be seen as almost synonymous to the need to be able to do the desired task in the services world for consumers. **Basically, they both want and desire for things to just work**.

The following table demonstrates some of the key differences I have noticed between the organizations in the internal world, and organizations in the external world. Interestingly, when an organization that operates in the external world are not as mature in the UX space, or have limited knowledge, their characteristics are more similar to the internal world.

This just amplifies the point, when you start out remember to **start lean with a low overhead, move in a rapid, agile manner and prove the value of UX.**

'Internal' – Employee Facing Functions & Less Mature Consumer Facing Businesses	'External' – Consumer Facing Products & More Mature Internal Functions
Internal politics lead to decisions that can detriment the experience as UX is not seen as important as other functions.	Usually experience is the key for making sales and customer loyalty and satisfaction.
Funds usually lower to spend on UX practices and work efforts.	Funds higher as experience equates to sales.
Higher time constraints as experience work is not normally factored in.	Set time to learn and test before going to market. Market research is a must.
Leadership and organizational alignment not always savvy in experience fundamentals.	Usually have teams or leaders responsible for the experience.
Enabling employees to do their job is key to their experience - they are somewhat captive to the tools and products they are given.	The customer normally has choices on where to spend money. Therefore things such as aesthetics can differentiate their choice even more so.

Table 1: Differences in internal and external organizations

A WORD ON SOME OF THE EXISTING MODELS

As previously mentioned, the 20th century focus of automation and efficiency has been flipped over given changes such as social media and mobility, giving way to concentration on the human element to experience, and human relationships. Before, automation led to efficiency and productivity and now, collaboration is seen as a key productivity factor. Humanization in organizations is here, and must be addressed if a company wants to successfully design experiences.

Previously, 'vanilla' was strived for, to cut costs and streamline process, it was literally 'the more standardized the better'. However now a balance must be struck and vanilla no longer works. Users feel the need to be treated as a 'segment of one', have a personalized experience, maintain control and be able to customize their experience. It is not feasible for companies to treat everyone as a complete individual, let's not kid ourselves. However, the key is that the user should feel like a 'segment of one'. Therefore in this new world, communications and marketing play a pivotal role in shaping the experience. The fact is, given the rise in social, a brand can no longer be 'managed' per-se. **The brand can only be born and cultivated through its value and through its demonstration of that value**. This must be based on the individual's need, causing more transparency than ever to become vital. Experiences must be consistent and continuous - not invisible, but visibly present, acceptable and transparent.

This does not mean to say the fundamentals from existing theories are not to be carried through into this new world, in fact quite the contrary. Human behavior is still based on the same thinking as before, just the behavior has now been augmented by a world of new and enhanced expectations as well as an **emphasis on emotion playing a key role.**

Design and usability are also evolving as the world does. A Google search will bring back thousands upon thousands of results with rules for design and usability. **The key for experience is to design for the goal and purpose, and usability is ensuring that the user can do their task easily**. One name that will come up on any usability search is that of Jakob Neilson. Many say he defined usability, particularly to the informational website space, and he has many great resources in the domain. His theory is based on cognitive science, and therefore can be combined with other theories such as Donald Norman's theory on emotional design, to deal with this new era of design. It is important to realize, his fundamental theories on usability such as 'usability is defined as supporting the users task, i.e. making it easy for the user to do what they want to do' are key, especially in a task based world. However, with the technological advances and new social aspects, they need to be augmented to include other considerations all the way through to delivery. What you are designing and the experience of it, should dictate the design e.g. a marketing poster would be completely different in design to an informational one of how to operate a badge card reader - however core concepts pertain to the design of both e.g. the message should be clear without causing the user frustration. He also states that products that are usable should be able to be used with little or no explanation - this is clearly applicable and a great theory to keep in mind.

The key here is knowing which theories to apply given what you are designing. This book will include references to many of these great theories that defined and shaped the UX world.

WHAT THE NEW WORLD BRINGS US

The 21st century brings us several themes that must be taken into account in each stage of designing the experience.

Emotions

As previously mentioned, the emotional aspect that was stripped in the last century, has become extremely critical in the success of designing experiences. There are two complementing theories on why we as humans make emotional connections -

- **The Action Theory** states emotions are hard wired into the deepest part of our brain - indicating, if a pattern of ideas is coherent to us, we feel calm, and when the pieces do not fit together we feel tension.
- **The Experience Model** states the experience is tied to distinctive and personally meaningful episodes we encounter - this draws attention to the fact that each experience you give a person will shape their emotion.

Design can therefore evoke emotions positively, if done correctly, and it has become a must that experience is emotionally meaningful.

Visibly Great, Not Invisible

It is often thought that in an organization services should be as invisible as possible to the users, so that they can concentrate on their other tasks. For example, computer maintenance should execute quietly and the user should never know. While automation should be used where it makes sense, it is important to understand that experience can rarely be truly invisible. The moment a customer steps into a store, a user connects to a network, uses a PC, drinks the provided coffee - it is a part of their experience. It is intensely and absolutely visible at all times, therefore the experience needs to be designed, with careful conscious decisions made as to how and when you show up to the user; as the experience is any process we are aware of, interact with or are involved with.

Social Aspects

There are many reports on the usage trends of social networks. They grew wide and fast, they allowed people to find people, connect after years apart, see what others are doing, locate where friends are, and even predict people's behavior. It has gone as far as studies showing people feeling guilty when they are 'off the air'. This has huge impacts on experience. It increases the expectations they have of tools and services, especially those related to collaboration. It also affects their virtual working environment for those that work in global teams. This effect is often spoken of, however, the other effect is one extremely important to take into consideration. This new world of social and mobile living, has caused people to experience information overload on a whole new level.

They are exhausted by being constantly on. For example, the sheer amount of emails a user gets to their personal email fatigues them, and affects their likelihood to read emails in their work inbox unless they are seen to be important. The other effect is the fact that given things such as having personal email and work email on the same device, Facebook and YouTube just an URL or App away, the lines between work and life have been blurred. You could argue there is no more work life balance, there is just life. This is a key to experience, allowing users to experience life, of which work is a part, is an inevitable byproduct of the world of today.

Nation of Now - Expectations

A key element, tied to the notion of time, is that we have become a 'Nation of Now'. We no longer have to leave the house to buy products, we do not have to go to the library to search for answers, we do not have to go to the shop and purchase the song we want to listen to...we get everything, right now, wherever we are. We want an answer, we go to Google. We want a song, we go to iTunes, and a movie, Netflix, we expect things to be obtainable, and fast. This has huge impact on experience. We do not expect the search function to not give the answer we search for, we do not expect to have difficulty ordering items we need, nor do we expect our actions to take time. While this new world has come upon us, and many organizations are still playing catch up - it is abundantly clear that people still prefer simplicity. Just because the technology has become more complex, people are not expecting what they experience to be more complex. In fact just the contrary. This faster world of instant gratification has upped the stakes for consumer experience. Since people now have easier access to so many products and services, experience has become the differentiator, and that is the same regardless within which world you operate.

Services Mindset

One way to start internalizing the new world and what it means to experience, is to think of everything provided to users as 'services'. With that comes the same concepts of service design as seen in consumer services and hence why the guidelines in this book can be applied to the design of anything. More importantly, understanding that the impressions of a service brand comes from the encounter with the service, as well as every interaction following. This is key, as **experience is affected by everything the user interacts with**. This has moved even one step further, into experiential services - services that are designed to focus on the interaction with the organization, as opposed to just functional benefits. For example, take soft drink company C, if they see the drink as a service – how would that change the experience they deliver? It would help them connect the dots between the actual service; the users interaction with the service, the support of the service...you get the picture.

WHAT THINGS DO THESE FUNDAMENTALS PERTAIN TO?

The fundamentals in this book can be used for basically everything that is designed for a user. This can be services, a marketing campaign, a workspace or a process. The guidance in this book can be used to design the experience of anything, at any level of granularity, be it a single communication, a process, a launch event, or a whole new application. This concept of 'anything' is referred to as the 'solution' you are designing throughout this book.

Here are some examples from various organizations.

Organization	Example Solution Being Designed
Workplace Design	Building a new workspace
Human Resources	Building a career development tool
IT	Releasing a new service
Marketing	Creating a new marketing strategy to boost morale
Hotel Service Department	Guest check-in experience

Table 2: Example solutions that could benefit from the framework

ONWARDS....

Now let's start with some of these fundamentals. Remember, the way in which you execute, and the level of pertinence of each section, will be dependent on your organization and where they are on their internalization and execution of focusing on user experience design.

We will define some key concepts that will be referred to throughout the book. Together they are nuggets of information creating the foundation to understanding how to design experiences.

2

KEY CONCEPTS

EXPERIENCE VS. USABILITY

It seems everywhere I go with new organizations that are starting to work more on user experience, the terms experience and usability are used interchangeably or they are confused. I want to start by explaining some of the simple differences between them, as they will be used throughout this book.

Think of experience as a whole space. It includes everything the person using the solution feels, sees, hears and touches. It is in the moment they use the solution itself, everything they encountered beforehand and everything afterwards.

Now, think of usability as a more narrowly focused term. It looks at the ease in which the person can complete the desired task.

Then it is easy to see, that to achieve a desirable experience, one must look at many factors of which the usability of all individual components making up the experience must be achieved. Usability itself can therefore be described as a core constituent of experience.

Let's illustrate these terms with an example.

Let's take the Human Resources team, who are creating a new mentoring system within the organization.

The experience of the system to a mentee using it will include things such as:

- Everything they have heard about the success and failures of the system from other colleagues and friends.
- Knowledge they may have of available mentors and their perceived quality in terms of what they can learn from them.
- The ease in which they can use the system itself to find an appropriate mentor.
- The amount to which the system meets their mentee needs.
- The support they get when they don't know how to use the system.
- The benefits they get on an ongoing basis from the system, in terms of enriching their career and intended value of use.

The usability of the system will refer to things such as:

- How easily can the mentee find a mentor?
- How easily can the mentee register as a mentee?

The space of experience can have many dimensions, with many elements that individually require high usability.

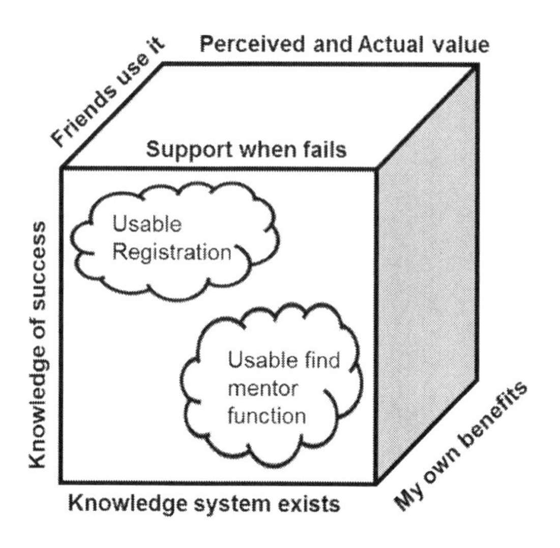

Figure 1: Example visualization of the space of experience, with elements requiring high usability

You can see how the experience of the solution and the satisfaction the user gets from the solution, depends on more things over a period of time, whereas the usability comes into play for each task they must do at a specific time.

WHAT MAKES UP THE EXPERIENCE?

To merely state that experience is a vast space with many variables effecting it is not that useful in terms of what you can implement. However, it is a key piece of understanding that needs to be internalized by not just the person implementing experience-based projects, but by everyone involved, including managers, executives, designers and communicators. Many times they may not fully understand the scope of UX.

There are two types of factors that affect the user's experience at a more concrete level that I have termed foundational and elements:

- **Foundational Factors:** those things that must feed into projects and form the basis of the experience for the solution being designed.
- **Elements:** those things that affect the experience the person will have with the solution itself.

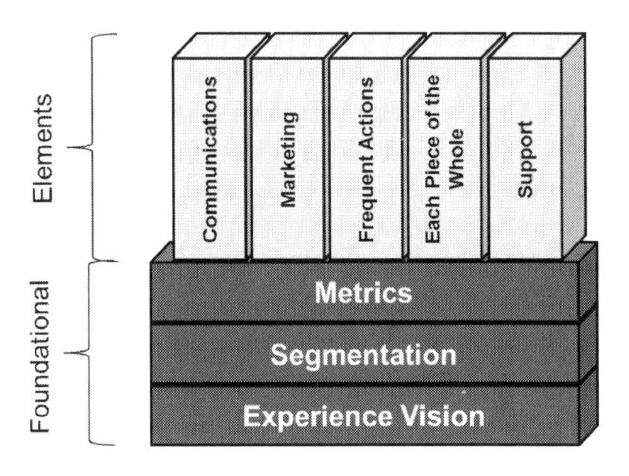

Figure 2 : The different factors composing of experience

The foundational factors compose of a strong base in understanding your users and your experience strategy. These are items that need to be carefully thought out and that can then filter throughout all of the elements.

Experience Vision: What is the one thing that is the ultimate goal for the end user? Be warned, this does not mean that you list all the features of the solution, which is a common mistake people make when trying to define their experience vision. The vision is also one of the most important factors as it trickles into every part that follows, including the design and measurements. Think of it almost the same as a business' mission statement, but this is the end user's mission statement, sometimes referred to as the **User Goal**.

Example:

Solution being designed: A Virtual Meeting Tool.

Experience Vision: Meet your colleagues and clients face to face, anytime, anywhere, without leaving your chair.

What the vision is not: Allowing end users to make calls, share documents and co-edit a shared space via online systems.

Why limit the potential of an amazing experience with a list of features that may or may not change in the future? In the experience vision, it is independent of technology, infrastructure and features; it solely pertains to the vision of what the virtual meeting system will allow the user to experience.

What is important, is the special words that states what the system value **is** to the user. It will allow users to use it anywhere, which means they could be on the move or stationary; anytime, so it must be always on; and face to face, indicating some form of seeing the other person. This could be now, via a webcam, or in 2050 when virtual reality meetings superimpose the person into the space in front of you. The experience vision for the value of this system to a user of it remains, regardless of implementation. It goes without saying that experience visions may change depending if the solution strategy needs to change given market or situation changes, however a new vision would be written in the same way.

Segmentation: Who are your users, defined for the purpose of the solution you are designing the experience for? What variables affect their behavior? There should be relevant and current data on your user base, allowing you to almost predict their behavior, as well as know how to manipulate it. Yes, I use the word manipulate, not in a negative manner, but more as to describe the fact that a part of the trick is getting the users to adhere and take on the desired behavior. This should feed into the elements to ensure the design, communications etc. are centered on good knowledge of who the users are. For this, studies should focus on what each segments needs are, their pain points, their typical behaviors, and how likely they are to adopt. Segmentation data can be composed of multiple input streams, and creating personas for your segmentation is very valuable in experience design.

Personas are fictional characters that represent the different types of users that may interact with your experience. They can be used throughout the lifecycle, for design, in marketing, in testing etc.

Metrics: This is a huge part of the business world where numbers are everything. How the success of the solution will be measured, what is measured and how, must be taken into account throughout each of the elements. And no, this does not mean asking 2000 end users if they are happy or satisfied in a yes/no survey question. In fact, there has been some work on the magic 400. Basically it states that if your test set is representative and if your total population is huge, then at even just 400 responses you can see the relevant patterns and that there is no real reason to need more surveys. However, how you measure will be dependent on the data and goal of testing. For many measures, there are more insightful methods that require less users. We cover more on this in testing (Chapter 6).

The elements are of equal importance in achieving a great experience for which ever solution you are designing.

Communications: These need to be carefully thought about throughout the process, they need to be specially designed and integrated to achieve the required outcome (remember they are a core element of the experience of the solution given the experience will be everything the person hears and sees regarding the solution). What a user sees and hears via communications is fundamental in driving adoption of a solution, that, and the fact the solution should be a truly better way to do their task, of course.

Marketing: The way in which the solution is marketed is very important, especially now, when what you say can be checked in an instant by peers and other users. Before, a commercial or marketing campaign detailed to users the drive to purchase solutions – now what you say can be exposed to hundreds of reviews all accessible in an instant. Sometimes glitz, glam and 'sex appeal' are still needed and can be very effective in marketing a solution, but be careful as sometimes this can be seen as annoying and pointless. It all depends on the goal, 'climate' and user base. This goes hand in hand with communications, to help drive adoption of the solution.

Frequent Actions: Parts of the solution that are used frequently, must have a great experience. The fact they are used frequently means they will weigh heavily on the overall experience if they are bad, as the user will be exposed to the bad experience multiple times. However, this being said, if they are good it doesn't necessarily mean the overall experience is good - it could just be an expectation of the user that the certain piece is what they qualify as 'good'. For this, usability testing can be used to ensure the action at that defined moment is usable. For example, think about getting a glass of water from the tap every day, multiple times. If our experience was bad, like the pressure was irate and you got splashed, or the water merely trickled and took ages to fill up your cup, this would really frustrate you. However, if the act of getting the cup of water was seamless, you probably wouldn't be overly ecstatic about it and say how wonderful it was that you could get water so easily.

Each Piece of the Whole: Each piece that makes up the total solution, should be at least usable to the point that the person can complete the task needed. This doesn't mean to say, given the balance between money and time, that they are always amazing, however, they do need to be of acceptable usability - which usability and other types of user testing can tell you. You need to pick the right times to wow users and know when certain functions and features are ok for a user without the wow factor.

Support: When a person gets stuck or something goes wrong, it should be easy to fix and continue. Perhaps automated with the person not even knowing, or perhaps guided fixes that help them fix it themselves, or even just a known way out - a number to call where a friendly and knowledgeable person awaits to help them. Either which way, it should be simple and quick to fix the problem and continue, the person should never be in a situation where they do not know what to do. It is argued, and rightly so, the support experience element is one of the most important ones, as it can salvage the total experience. We have all had the product (laptop or router perhaps) that broke, and the experience we got with the support process and staff or tools either made us dislike the product immensely, or really like them and tell everyone how amazing it was, even if the product itself was somewhat mediocre. The reason is simple, in a moment of frustration and disappointment (with the solution), you have the opportunity to wow the person, to make everything they are experiencing that is bad, good again, and this plays a heavy weight on our minds and memories.

This has been a lot of information to internalize, and, given its importance, let's illustrate it with a few examples to show how the elements fit together to create an end-to-end experience.

	A Computer	A Break Room
Communications	What does the person read about the computer? Startup guide, the renewal notice for maintenance etc.	What has the person heard of the break room? What has been communicated as to awareness?
Marketing	What advertisements have they seen for the computer? What expectations has this led them to have? What reviews exist?	What subtle marketing has occurred so that the person sees the break room as a perk, can find it and appreciates the value of it?
Frequent Actions	Switching the computer on, switching it off, connecting a peripheral....these things occur frequently during use of the computer and therefore imperatively must not cause annoyance and must act as expected.	Impromptu conversation (can lead to a better social environment and higher innovation), purchasing from vending machine, getting coffee. All these must be easy and natural in occurrence.

Each Piece of the Whole	The monitor, the keyboard, the ports...must all individually be usable.	Number of seats adequate for use, position of seats and tables, comfort of seats, flow of room, vending machine stock...
Support	The computer hangs at start up, how easily is the situation rectified and with what quality of service?	The vending machine breaks...what does the user do?

Table 3: Examples of the elements

These pieces combined, will allow focus on the necessary factors that affect the overall experience. When all are achieved, the thought process leads to a total, great experience.

ROAD™ TO A GREAT EXPERIENCE

There are four main phases that the design of anything should go through, think of this as the overall framework, or structure. A way to think through the experience you are designing.

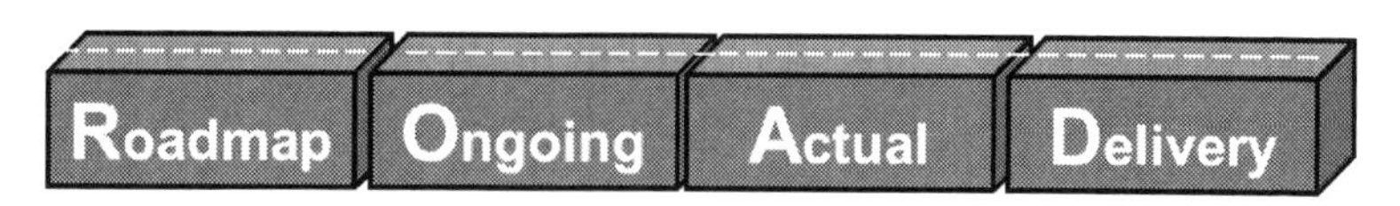

Figure 3: Four phases of the ROAD™ framework

Roadmap: This refers to the strategic direction in which you are taking the solution that you are designing. It looks at what the state the current solution is in, where it is going now and in the future. Or, if nothing exists today, then look at how users carry out the task today and learn from that.

Ongoing: This refers to the ongoing plans for the solution you are designing. It looks at ongoing activities that will need to be implemented to ensure success and growth. This needs consideration as early as possible, and will continue to grow throughout the solutions lifespan. There is a reason Ongoing is second (not just because it makes an easy to remember acronym), that's because there should be some early thought as to ongoing success and strategy to enable a great design and delivery. Many people think of their ongoing experience measures and feedback/feed-forward loops and mechanisms last…this can lead to failure or a longer time to reach success.

Actual: This refers to the actual solution you are designing, be it a product, tool, service, campaign or process. It looks at the elements of the actual solution and how they fit together. It also looks at the actual design of the solution. We do not focus on visual design in this book, as usually there is a UI team, a design team, an industrial designer or something similar that works on the visual design. The visual aspect of the design is a whole book more. Here, we concentrate on the design of the experience as 'design isn't just about making things beautiful, it is about making things work beautifully' (Roger L. Martin).

Delivery: This refers to the process of releasing the solution to the intended audience. It looks at how the solution is rolled out to ensure maximum satisfaction and levels of use to obtain the desired return on investment (ROI, basically getting the value needed to make the cost worthwhile).

Let's look at an example to illustrate this point concretely.

Taking the Facilities Department project of designing a new shared break room for their users, and putting it through the ROAD™ framework; they would consider the following types of things:

- *Roadmap:* Is this the final design for several years, or a phased design with quick iterations of improvements as budget becomes available? Who is it that will be using the space and what is the spaces purpose?
- *Ongoing:* Where and how can users leave their feedback and suggestions for the space? What happens if something in the space breaks such as the tables, vending machines etc.?
- *Actual:* How should they design the space itself to meet the goal of the space? Where should furnishings such as tables and chairs be placed? How many areas of seating should they have? How many power sockets?
- *Delivery:* What do the users do now that having this space will change for them? How will they

know about it? How will they use it in the intended fashion?

These are just some of the questions that the Facilities team would start asking themselves within the ROAD™ Framework, and illustrates how each piece links together. It is the consideration of, and answers to such questions that lead to a great experience design.

USER CLIMATE

This is a term I have come to use to describe the general sense and emotions 'in the air' so to speak, with regards to the user. As you read in Chapter One, the changes the world has brought us means the way users feel cannot be ignored, nor taken for granted, as it will indeed affect their behavior, response and general satisfaction. It needs to be taken into account at varying levels of granularity such as climate of the company itself, climate of the product space itself etc.

The main components, and why I think climate is such a good term, are:

Typical climate and patterns seen by variables exist, that can be considered the norm.

For example, just as England (variable = location), in December (variable = month) is cold, you can look at the typical feelings and behaviors of user groups. For example, users in China (variable = location) are typically less likely to rate a satisfaction survey poorly, or users in sales (variable = function) at end of year (+ variable = time) are usually stressed.

Daily weather is dependent on current variables.

This can change daily depending on what is currently happening, for example, a cold front blows in for a few days, so weather is chillier than usual. For example, an individual could have a bad day, their PC can break, they got up early etc....can all effect their current emotion.

Ability to forecast.

Given the norms, and the current outlook, there is an ability to forecast responses and feelings. For example if the cold front is not leaving for at least 3 days, then we know the next 3 days will be colder than usual. In the same way, if we know its end of year and the next 2 weeks will be busy for the users, we can choose not to bombard them with surveys, or even proactively check that they have everything they need to stay productive especially in these 2 weeks. We can use what we know is coming up to create a better experience.

Random, unsuspected events that cause deviation from the norm.

Just as a random thunderstorm or tornado can occur, users can have random events that they were not prepared for. For example, employees can go through leadership changes and reorganizations, large layoffs, pay cuts etc., all can happen at an unplanned moment, and can change the climate immediately, causing a certain level of sensitivity to be needed.

INNOVATION VS. INCREMENTAL IMPROVEMENT

A key fundamental is knowing when there is a need to innovate vs. a need to continue to make incremental improvements. This will depend on the constraints you have and the balance between effort and increased gain in user satisfaction.

As you can see from the diagram, you could be on the right path of improvements to the best experience, or you could be on an improvement path that will only lead to a local maxima. The question is, is the local maxima enough, or do you have to innovate?

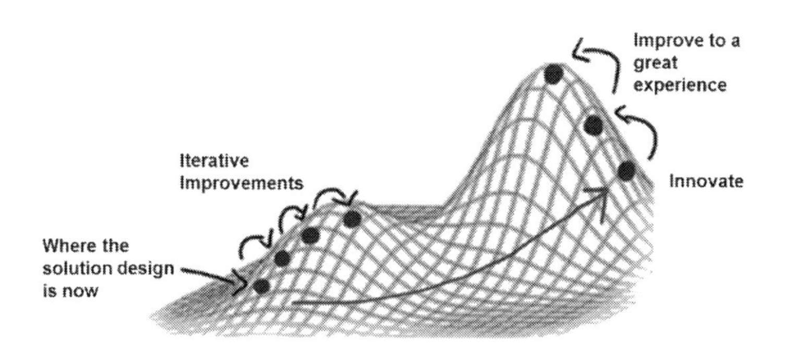

Figure 4: Improve or innovate

The truth is sometimes an innovation can differentiate a product in the market place, sometimes an innovation may be the only way to save the experience of the solution for the users, and other times simple improvements or 'picking the low hanging fruit' may be enough. When starting lean, it is good to know if the increment will give you enough improvement to the experience. Know how to weigh up what you focus on if you cannot focus on everything. Know what the couple of things are that will hook the user and keep them there.

I have found that you can only realize whether you need to innovate or improve by testing and analyzing at least 4 things:

- Impact of the improvement in terms of your goals.
- History of data on satisfaction and issues with the solution to date.
- Alternatives and testing those concepts against the current solution.
- Current restrictions including time, project resources and money.

Sometimes, if the proposed improvement gained is not enough then there is no point to the project or work effort until you can innovate, and resources are better spent elsewhere. Innovate, means start again, look at the problem and think through how to design the solution again regardless of the current design. After you have rethought the solution and the best way to design it, if you can salvage pieces of the existing solution then of course do so. Improve, means look at what you can do given the current design as a starting point.

Let's look at an example:

For this example, Company X's employees have a big issue when they bring guests into the building as they cannot get them onto the internet without filling out a lengthy form and submitting it via a terminal at the reception desk multiple days in advance.

Improvement: When the guest signs in themselves at the terminal, the employee authenticates them and then provides them a code to enter generated on the fly and they are able to access the network. This would cut out the several steps for the employee.

Innovation: The guest takes a picture of the QR code in the building lobby; it directs them to a page where they self-authenticate entering a code on the back of their badge, which was generated at sign in. The employee has nothing to do, apart from register the guests visit online before the visit occurs.

While the innovation may be cool, perhaps cutting down the steps is enough for the employee to be fine with the process. This improvement takes minor work and hardly any cost, therefore, improvement should be the chosen option.

Looking at the history is a great way to see if you are likely to have to innovate. If each year there has been an improvement project fixing minor issues, and each year the users remain dissatisfied with the solution, then it is likely you are either fixing the wrong problem, or the foundation is so broken that incrementally fixing it will not help, and innovation is necessary.

It is worth noting that there are different ways to think of innovation. It can be a huge change in dynamic, or it can also be a much better way of doing something through connecting the dots and removing effort. For example, let's take the need of a person to get to a location via car without driving themselves. Innovation could be a self-driving car – this is a huge change in dynamic that requires a high level of change. Innovation could also be an application like 'Uber' which takes out the pains of ordering a taxi, such as enabling the ability to order from your phone and pay with credit card automatically etc.

DATA

Data, is something that is incredibly important. It will play a part in almost every stage of the ROAD™ framework and is the basis of most decisions. It is also incredibly dangerous. Misinterpretation and misunderstanding can lead to whole initiatives, products and projects not delivering the promised value. Understanding the core points on data is key, as without a doubt, leadership will want to see numbers on everything you do. Part of your work in the space is to help leadership and management understand what data means in the space of experience, as more often than not, they will need to internalize these concepts also. From the successful organizations I have heard speak to the topic, there were three key trends that were at the core of their success.

Rule One: Right people, right data, right time.

Seems so obvious, yet it's shocking how hard this is to see working correctly in organizations that are new to the space, or ones that simply don't know what this means or even more scarily, think they do have all three right.

Right people – The analyzing of experience data, the interpretation of how people think and behave, should be left to professionals of the space. While there are some patterns, and the skill can be taught, it is fundamental to have someone that understands this data type leading the various data capture and analysis activities. Sometimes what happens in the organization is the executive wants to see what users think, they set the task of having a focus group to whomever on their team has the time and capacity to do it. This person has no idea how to run a valid focus group, or how to interpret the data, therefore the findings can be invalid.

Right people, also refers to having the right people in your participant set. It is not as simple as just getting 2000 responses and being statistically relevant, it's about understanding the group you need to learn from and who you are designing the experience for. It doesn't take a genius to figure out if you test a bicycle for 5 year old boys with 10 year old girls, they will likely act completely different. Yet, in the corporate world, sometimes people are clouded by the fact they feel as many 'responses' as possible are needed, with little regard as to who is responding. We will delve more into testing types and how to conduct them in Chapter 6.

Right data - Alarmingly, much of the data used is wrong. Assumptions have been made or even grand extrapolations that lead to wrong decisions. For example, someone could read that 70% of employees are dissatisfied and they know a few people that mentioned disliking the salaries they get. They then conclude that salary is a core dissatisfier. How did they know that employees wouldn't have valued comp time, or been ok with an alternative plan? Another key trend I see, is people listening to the wants. The key is to investigate the needs, and deliver a solution that is tested to work for the target audience. Listening to what a user says they 'want', will lead to a vicious cycle, as wants can vary and depend on climate, and **after one want is met, there is usually another want**.

This is a key point that I want to illustrate with an example. A company looks at the comments of a satisfaction survey, and identifies a trend of people asking for iPhones. This is a want. They do some concept testing with some focus groups to investigate further. They find that people actually wanted the ability to do emails for work on the go, and actually it was only really the sales people. They tailored a cost effective solution that dealt with the need of sales people to work on the go. Really they had just said iPhones as it was the first phone name that came to mind.

Right time - There is no point designing something, releasing it, then seeing it doesn't achieve the value or experience, and having to rework it. Use the right test at the right time. Common sense...I know, but it truthfully happens in so many organizations and projects. They are up against hard timelines and just execute. Then when it turns out to be wrong they end up feeding this wrong data into endless iteration cycles. More of this in testing.

Rule Two: Don't be Afraid to Throw Away Data.

Organizations are drowning in data. When you come to using data in designing the experience, look at what is relevant (right data), and don't be afraid to disregard the rest. Just because you have it, or time was already spent gathering it, it doesn't matter...if it's wrong or invalid then don't use it. This sounds simple, but you rarely see organizations disregard data, they tend to succumb to the 'we have it so we must use it' phenomenon.

Rule Three: Don't Assume You Need New Data.

Another trend we see a lot of from organizations that have gone through a history causing user dissatisfaction, is the fact they have a lot of change at the management levels, or they have several iterative projects all trying time after time to do the same thing. What happens is, they have data that tells them everything they need, but the fact that it is a new leader or new project leads them to think they need to capture more data. This is a big waste of time. Only add to the data pool when you have to. Trends show you a lot and sometimes it is only the case that you need some specific focused data or even no additional data at all.

PERCEPTION OF TIME

Time is something that no one has much of and certainly people have none to waste. However there are some trends seen internally that are more heightened with respect to time than even in the external or consumer world. Time is something to always take into consideration when designing any solution. Even a marketing campaign for example, can take too long to understand. In the situation where you are asking the users to do some action, time becomes even more important. As previously mentioned, we are seeing a rise in a user's need for instant gratification. This means it becomes even more important to set their expectations so they are not left wondering, and know what they can expect from a solution. It also means you should get to the point...fast. The point of your materials, the point of your pitch, your application...you get the drift. **Users should see some value or return very quickly**.

Perception of Time – Enterprise Users

It is very interesting, in the internal world, employees seem to have not 30 seconds to glance through an email and certainly not 10min to install something or fill out a form...yet, they have plenty of minutes in the day to peruse YouTube or chatter idly on the internal IM. Employees believe they have absolutely no time to do what you ask, therefore it is imperative even more so that they understand why they are doing something. They need to see a point to the task which is pertinent to them and that the act is as short as expected. This is heavily linked to the work-life clash, where the day has morphed into a longer than 8 hour day, sprinkled with work and life activities. More important for you to understand is the thought behind the activities. The employee has chosen to look at that 3min YouTube video for a break, or for whatever reason. It was their choice, therefore making up those extra minutes of work isn't even an issue to them, as it was their own choice to do so. The organization asking them to do something was not their choice, and the time they spend doing it is therefore an irritant given they will have to make up the work time.

You can relay this concept to almost any user in the world today. Think to your own experiences, when you get frustrated because an installation takes too long, or a download is hanging and you do not know why.

LINK TO WORK/LIFE PRODUCTIVITY

This is not my job syndrome also leads to dissatisfaction. Users want to spend their x hours a day being productive in their own jobs and living their lives using their time on things important to them. Extras and things that are perceived as taking too long, detract from this and annoy them easily. The more time they waste on things they don't feel they should have to do, the more they dislike it. Do not just think of productivity in terms of work, think of it as life productivity also. When users haven't directly chosen to spend time using your application or service, they will like you more when you use their time wisely and they feel it is useful and pertinent so that they can continue being productive in their life – run those errands, spend time with family or friends etc.

HOW DO THESE PIECES FIT TOGETHER?

Given all the fundamental concepts you have read so far, it may be hard to see how all the pieces fit together. The diagram in figure 5 depicts how the factors that affect the experience fit into the ROAD™ framework. The other fundamentals were concepts and key learnings to keep in your mind throughout the remaining chapters. Which data and when the data is used to make decisions will depend on the moment. For example, data in the roadmap phase could be data on the current experience users are having and data in the ongoing phase could be satisfaction scores

Think of it as a way to execute, what to execute on, and how to execute (the principles coming up). We will build out the rest of the components throughout the book.

It is worth mentioning that these principals and the thought process can work in whichever design-development lifecycle you are following. You can see how it fits into traditional waterfall lifecycles and also a more agile process. For agile, you would ensure the strategic UX work is completed such as personas and UX vision at the beginning. You would then move into fast iterations or sprints of design-development-test-release with user flows and usability testing as you design more features.

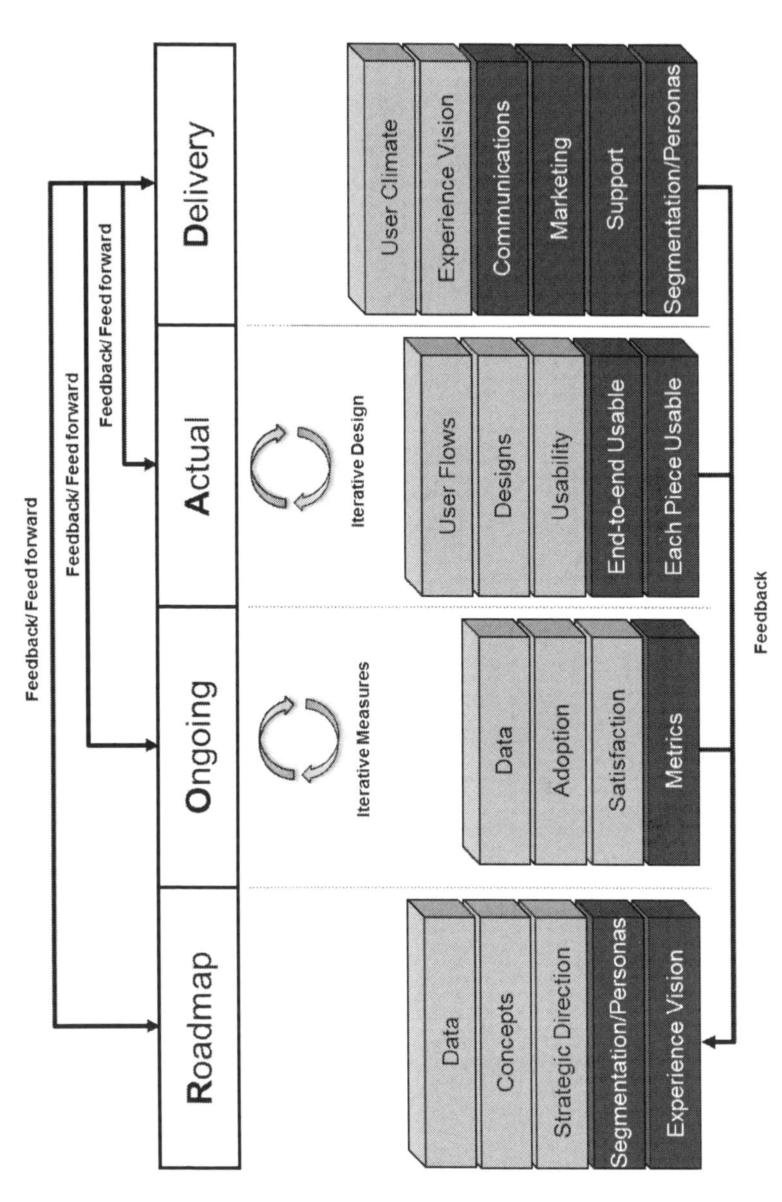

Figure 5: How the pieces fit together in the ROAD™ framework

3

EXPERIENCE IN THE STRATEGY STAGE

The strategy behind what you are doing is one of the first things you want to think about. Even sending a communication has a reason behind it and a desired outcome. Sometimes actions and projects in a corporation are not solely to improve the users' experience, but are necessary for business growth and can sometimes even be not what the user would want. Outsourcing support services is a common example. While outsourcing this work is cheaper for the business, the quality of the service provided may not be adequate as issues like language barriers may arise. You can argue that the experience can sometimes be about managing the change, by ensuring that users are not impacted, that their disruption is minimized during the change, and that they can appreciate the reason for the change. Sometimes with some careful thought, it is even possible to convert a potential bad situation into something beneficial for them.

There are some key basics you can do in this stage to ensure as good as an experience as possible. The work in this section can be maintained and carried out by a centralized team in a consultative fashion, however it is vital that the overall owner of the solution, whichever department they may lie in, is a part of the process. Ideally, they will have given thought to how to manage the change as part of their normal strategizing process. For an example to illustrate, let's take a company that sells collaboration solutions and their team responsible for their instant messaging solution. This team needs to create the next iteration of the tool with enhanced features. The strategy work is carried out by the company's experience team, who provides the recommendations on feature set and UI to the instant message product and development team to implement, or the instant message team have their own experience designer within the team who can complete the work in this section.

WHAT IS THE SCOPE OF THE SOLUTION BEING DESIGNED?

The first question to ask yourself is, 'what is the scope?' Is it a communication being written, a process, a product, a space...? What exactly is it you are designing? What is the solution?

Obviously, while the thought process to design the solution with a good experience is the same, the scope will help you determine the effort and time required in the roadmap stage. The key is to ensure enough time is planned for, or seeing how to fit in what amount of work in the case of a constrained timeline.

Time needed will also depend on the next point, if you have data already or need to collect some, and if so how much you need will depend on the scope. The very high level example below shows considerations for the difference between a communication being designed, and a new coffee experience in a break room.

	A Communication	New Coffee Machine
Scope	Content written and approved.	Machine design, prototype development, testing and deployment.
Data	Audience, message purpose.	User base, features, current experience gaps.
Data Needed	What the changes are in relation to user type.	Current experience evaluation of current coffee machine. Concept testing.
Time	3 weeks until the send date. 1 week for data collection on the impact of the change. 1 week for content creation and approvals. 1 week buffer.	8 months to go-to-market. 1 month research and ideation – experience work required in this timeline including, usability testing, change impact analysis and deployment planning.

Table 4: Different scope and considerations dependent on the solution

WHAT DATA DO YOU HAVE THAT SHOWS PURPOSE?

You need to look into **what the user value truly is**. This is the only way to know whether the experience is more about ensuring the minimization of disruption, if it's about creating a wow type experience, or perhaps, it is so destructive to the experience that another option needs to be considered.

Types of input data can be:

- Analysis of feedback or user testing on the current solution you are improving - if you have a current solution in place.
- Industry standards and benchmarks.
- Segmentation needs - the needs of different user types.
- Reports from metrics and measurements on the solution if it is existing.
- What there is today and how people use it? How will you make it easier, faster or better?
- What problems exist today?

You may not have any data to go on, and will need to collect data to see what the user value is. You may already have some data, and may need to validate it by carrying out a focus group, or some observation to see what the real value would be. Remember the data fundamentals here from Chapter 2, as the wrong data can lead to the wrong decisions!

In this stage, it is important to think through a couple more questions that maybe do not spring to mind as the usual input data.

HAVE POTENTIAL PROBLEMS BEEN RATIONALIZED BY THE TEAM, OR LEADERSHIP?

Whenever you hear the words 'yes, but' be cautious. Usually this means people are rationalizing the problem as opposed to getting to the core need and driving a solution. For example:

UX Person: 'we implemented a pass code for security and people hated it as they never remembered it and have to spend time calling support to reset it...'

Security: 'yes, but we had to as it wasn't secure before'.

UX Person: 'The business need was a pass code for security, the users need was an easy way to know what password to set up, and how to reset it easily if they forgot in a manner that didn't require calling.'

Security: 'Yes, but we couldn't implement a solution to reset it easily so we just gave the one we have'

UX Person: 'why not? You couldn't at least provide decent instructions or pick a better reset solution at the same cost?'

You get the gist. Now, in this stage is the exact correct time to look at the need, the data, the end users, and the options, and then bring the evidence to the team as to risks, options and recommendations. Do not just give theories back to the 'yes, buts', this is an important stage in the internalization process of user experience. It is important that you give them their choice points, their options and the impact of choosing each option on experience, cost and time at a minimum.

ARE PEOPLE ASSUMING END USER BEHAVIOURS?

Very much linked to 'yes, buts' is the tendency of people new to the area of user experience, or leadership and management to say 'a lot of people say' or 'I hear'...who said? Unfortunately, this is usually synonymous to 'I think' or 'I believe', you should not take that as a given, nor assume correctness. You should definitely use that as something to look for evidence of in your input data or initial testing. This is one case of 'guilty until proven innocent'. It happens more that you think, they hear one comment and believe it is the norm, or they go off their own experience and project that through a generalized statement as they believe 'I am a user, therefore I know what a user wants'. By following an experience design process your compass will be set to follow what your true value is to your target users. You will be able to ensure that your solution is responding accurately to the user's needs and fixing their actual problems. Knowing your experience vision and user goal will allow you stay grounded and focused on what you are trying to achieve.

EXPECTATIONS AND CLIMATE

Now is a good time to consider these two fundamental concepts that were previously explained. What do users expect from this solution given what they use outside of work, inside work and see in their lives? Will you meet expectations, or need to set the expectations so that they are not disappointed?

What is the climate? How do users feel about their current situation with the solution being designed? How does current morale or perception affect what you are about to design and deliver to them, will it likely be seen as good, bad or indifferent? To find out you can do a pulse check or just simply some interviews – **the key is actually talking to people and finding out.**

INNOVATE OR IMPROVE

Linked to the above is the question of all your options. If you have something existing will an improvement be enough, or is it fundamentally broken and therefore an improvement will not help and innovation is required. You can only really learn this by doing some actual end user testing, or concept testing on your ideas (Chapter 6). Remember, it could be the case that an improvement is enough, and your idea of what you think is needed is in fact grander than what is actually needed.

An example to illustrate this is - facilities have feedback that employees really hate the provided coffee, so they are implementing a new coffee station. They immediately think of barista machines etc. that would be fancy for the employees. However in concept testing, the employees while liking the idea of the machine, do not want the hassle of a barista machine unless a barista is there to make the coffee. Instead they actually just wanted a change in the brand of coffee used and creamer pouches instead of powder. The barista machine would have actually annoyed them more, and the company saved money as the equipment and ongoing cost of the barista machine was much higher than keeping the same brewing machine and changing the coffee and creamer provided. This example may seem trivial, however it demonstrates the point well. Not to mention good coffee has a remarkable impact on experience it would seem. Why? Because coffee is seen as a social element. The 5 minute coffee break with a colleague appeals to the employees' social emotional need, and also creates conversation that leads to a connected and creative work environment. I've even found that one of the first comments new employees make to their friends and coworkers, is the state of the free coffee provided. Just think how many times you may have heard someone comment on the coffee where you work....

Now I want to, at a high level, consolidate this step with an example to illustrate everything tied together.

Scenario:

An IT department decides that users should no longer be able to store PST files on networked shared drives (PST files are archived emails from their mailbox). The reason is purely because of cost cutting for IT. Reports had shown these files accounted for huge amounts of data storage and network bandwidth. They plan to email all users to state they will be deleting any PSTs stored over the next 14 days, as well as disabling PSTs. They think they have considered the experience of the solution as they are giving users fair warning.

What occurs?

The project team rightly looks at their data. Their segmentation data shows that PSTs are a huge part of employee's lives, especially the non-client facing ones. They warn management of the potential negative impact. Management responds with the fact that a lot of people, if they couldn't use PST's would be fine, as most PSTs are not even needed and most users wouldn't be impacted.

This rationalization and opinion from management is a red flag, so to avoid unintended consequences, the team does some further testing to add to their input data. They observe some users and how they use PSTs. They find many people were using it to store emails with attachments. They also do a quick spot check with a focus group on the concept of not using PSTs. They learn there would be panic and employees would really dislike it unless there was some alternative solution.

They also look at the other important factors. The core employee need was that people needed a place to store files they think are important as they are scared to lose any data they feel they may need at a later date. The current climate was that companywide moral was low after a series of cuts in employee benefits and provisions, therefore there was a high chance this would be seen as yet something else the company had taken away. The climate specific to the area of PSTs was that they were incredibly important, as employees thought they needed to save all emails in case they need evidence of something someone had said via email.

The expectations in this space were also going to play a large part. Many employees used Gmail, and Google gives masses of storage very easily. Therefore, employees are likely to wonder why they can't just save all their emails like they can in their personal life.

All combined, the user value of taking away PST's is limited and it is really more for business benefit so disruption must be minimized. A solution for helping employees to store their PST attachments and emails elsewhere must be considered. Would innovation be needed? The team considers it with their new cloud solution they are rolling out within the next year as a seamless email and cloud solution could be delivered, creating the chance to create a wow experience. Therefore they consider the option of holding the project to align with the other one.

After looking at feasibility, the project team need to stop PSTs before the cloud solution is rolled out, so they need to concentrate the experience on minimizing impact. They then move to consider the options they have, that are not just emailing employees the fact that PSTs will be deleted and disabled. They come up with a few ideas, company provided portable hard drives, limited PSTs which would need user management of their emails, user space for attachment saving, bigger hard drives, etc.

So for this team, their purpose is to stop network PST usage. In this case, the user's value of this solution will be limited, and the scale of the experience delivered will depend on which option they go with. The user need was needing a place to store files that they think are important. By going through this thought process, it enables you to think through potential consequences to different options and choices. It ensures you ask questions early on and then select the best way forward for the users and business, affording maximum adoption and therefore achievement of the proposed value.

WHAT IS YOUR USER EXPERIENCE GOAL?

Defining your experience goal is at the very core, as it will cascade into everything such as design, delivery, metrics etc. This is the just like a mission statement, it is everything you are standing up to. It should not be a list of features (remember the key point discussed in Chapter 2). From your initial data, you know what the users core need is, now think to the future also and ensure that your mission is translatable no matter what technology, tool or material you use. It is really the one thing the business wants the user to achieve with the solution.

For example, let us consider an HR career development tool. The need of users is to be able to navigate to the next step in their career within the company, this way the company can retain its talent via career growth.

The mission they come up with is - 'search for your next job easily using the online search tool and apply seamlessly using your online profile'. What's wrong with that?

Many things, but most importantly -

1. It specifies tool specifics such as online, but perhaps there will be a number to chat to someone in the future, they don't know that yet.
2. It specifies exact features of the system such as search, apply and profiles, which is pretty restrictive.
3. It uses the word easy - that is an absolute given, it should not exactly be hard now should it. Those words are also incredibly subjective.

How can you test your mission statement? One way is when making a decision about something in your solutions design can you say 'does that fit the mission statement?' and answer your question adequately.

In the above example, if we wanted to add a mentor section, we couldn't as we are limited to searching and applying. That's not to say user experience goals should not be flexible with world changes, but they should not be limited. If you strategically change direction, or solution strategy, then update your experience goal accordingly.

Think business mission statements. For example, a part of McDonald's mission statement is 'make every customer in every restaurant smile'. This is the one thing McDonald's wants people to do, and this can be cascaded into, quality of food, service experience, hours of operation and even consistency between restaurants. It literally is the key question to ask in their requirements, strategy decisions and design. For example - if many people in a university town want McDonald's at 11pm, and they close at 10pm would they smile? If they change the nugget chicken used, would they smile more than with the current nuggets?

Not getting this right is one of the main fail points for experience strategies. You should define this, before you make decisions on how or what to implement. All the data you gathered will help you as you start with why you are even doing this, what's the experience story? Again, this provides you a direction to stay true to.

Let's go back to our example of the HR career development tool. A better user experience goal could be 'connecting people to their next career move at Company X, where personal growth and development is always the next step, and we are there, every step of the way'. This allows the solution to be online, in person, a mobile app or anything, it allows any level of features as long as they are helping get the user to their next step. It also indicates value in growing and developing people. You can see how this can cascade into marketing and branding, for example a site slogan could be 'connect, grow, develop', or 'with you at every step in your path'. It also allows you to make decisions, for example should we enable people to apply direct to the recruiter? Yes, as it helps connect them and bring them a step closer. Should we send them job listings? No, as we state personal growth and development, so we need to personalize what we send them, perhaps making it just those jobs in their location, or just as an option for the user.

WHAT ARE YOUR CONSTRAINTS AND COMPETING GOALS?

There are always many goals and requirements to any project, such as business goals, as well as user experience goals. There are also usually many stakeholders, with their own respected interests in the solution.

It is important to understand what all of these are. Sometimes you can get this information from a proposal document, or a requirements document.

First, list out all your goals, and constraints e.g. goal = save business $x, constraint = cost. Other constraints include resources, time etc. Then see where experience actually lies. You may not be able to tell yourself from just this information and may need to find out by also talking to various sponsors or stakeholders, e.g. ask them what is important to them and what their goals are with the solution.

If there is no goal surrounding experience, then you create the goal (that's right, don't ask them for it, you create it), take it to them and see where it lies. It's important to get it on paper as a legitimate goal, in the correct place. Also, you need to check if everyone is in agreement and aligned to the goal. In more mature organizations, within reason, they may say if the experience is improved by that much more, they are willing to spend more.

It is worth mentioning here, you shouldn't think of the current design of the solution as a constraint, even if you are doing an 'improvement project'. The key is to start with an open mind as if you start out thinking you must start from what you have, then you will be at risk of not improving the experience enough as you may not entertain potential high impact improvements. Remember, even improvement has a scale.

WHERE DOES EXPERIENCE REALLY FALL WITHIN YOUR GOALS? AND DON'T LIE...

You do yourself no favors by 'pretending' that the experience of what you are designing is a top priority for you, the team, the project or the company. It is clear that sometimes people will say experience is important, when in fact it's more a checkbox or a buzz word. Sometimes, in reality, if it came down to a decision to increase project time to fix an experience issue, the answer would be no. This is usually the way in organizations less mature in the UX space, or in organizations where the leadership hasn't fully internalized the UX concepts and what it means. Sign's that may indicate that this is the case is if you see problems being rationalized, or if leadership tends to make decisions regarding experience based on what they themselves think. In this case, you will be far more successful if you understand where experience truly stands, and then design the best experience possible. This doesn't mean that you limit yourself, you can always try and ask for more, but at least you will plan for and be prepared for the true environment you are designing in i.e. recognize where experience lies in the list of priorities.

For example, if cost is the most important goal of the project, and experience is second, then you plan within the constraints. Going back to our coffee example, this doesn't mean you don't meet the user experience goal, you meet it with a good an experience as possible within the constraints. So here, the type of coffee was an issue, and great tasting coffee was the goal. Within the cost constraint, you want to make sure it's as great as possible, at least better and within cost. So this would affect your selection. Again, you would come up with the choices possible within the constraints, then test which one of those meet the experience goal.

A moment to recap, so far you have considered scope, the experience goal, user need, value and constraints. You have also looked at data to see what is really going on.

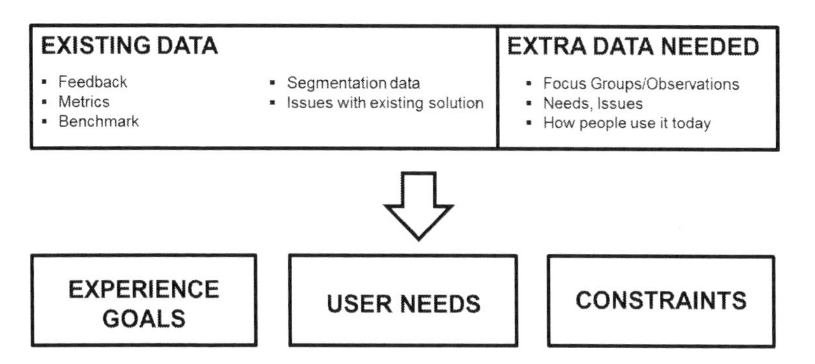

Figure 6: Data fed into the experience goals, user needs and constraints

PERSONAS

Looking at the users that will be using your solution is very important. You need to understand them in some detail to enable a design of an experience that works, as well as to help tailor marketing, content etc.

A persona is a representation of a user's demographics, goals, needs, environment, pain points, concerns, expectations and behavioral patterns. You can determine these by interviewing some users and then creating realistic representations of them in a document that the team can use for reference.

A typical persona consists of:

- Photo – to help visualize the character
- Fictional name
- Role
- Responsibilities
- Demographics – age, family, education etc.
- Pain points, concerns, issues
- Goals with the solution
- Environment
- Representative quote from the user

A quick internet search will bring you up several examples of personas if you need a closer look.

WHAT AFFECTS THE EXPERIENCE OF THE SOLUTION YOU ARE DESIGNING?

We have already described experience as a space. Given this, you can think of the variables that affect experience almost like a vector in the space, determining what the experience is.

You can see from the illustration in figure 7, again using the cube to illustrate the experience space, that where you are on the variable lines, determines where you are in the space.

Think of this as the key variables that will increase or decrease the experience for the solution you are designing. Think of them as knobs you can turn, and depending how much, will determine how good the experience is.

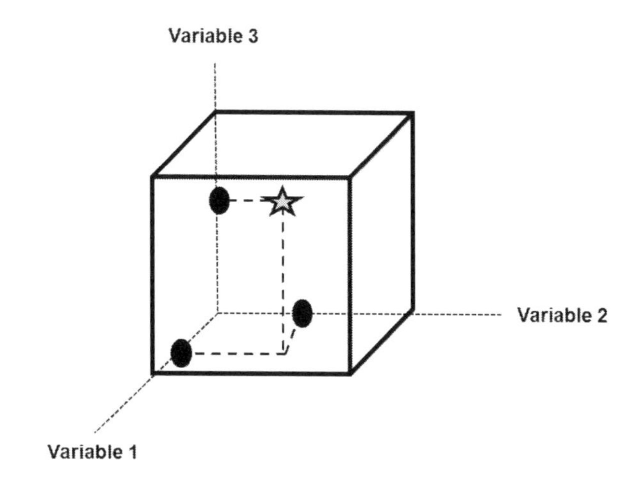

Figure 7: Experience variables

I have found that there are usually about 3-5 of these experience knobs that are vital to experience, which themselves can be broken down into sub-variables. You want to determine what drives are the strongest for the users. You can use the data such as comments in feedback to help determine them. Or if your data doesn't tell you, then find out by actually asking some users. Even if you determine these drivers from the data, then make sure you validate them as it is always best to not assume. When you do this, do not assume that what you think is right - for example do not tell the users what the variable is and then see if they agree. In determining the variables, start blank with questions that are very open about the area or solution. Then when you analyze the data, you can see which variables bubble up, in other words, what themes exist or are most mentioned.

Let's look at an example to help you see what we mean here. We will take the 'coffee experience' (yes, again).

The facilities team received a lot of feedback to their feedback box about the quality of the coffee. So quality is clearly a variable affecting the experience of the coffee for the employees. However, quality could mean anything, it's too vague to be a variable that they can stand up to. It could mean quality of coffee machine, coffee itself or condiments. They also have a few comments on availability, in terms of where it is situated in the office physically. They conduct a focus group to discuss the coffee and see what the users think of it.

Quality is one of the first things mentioned. 'The quality is disgusting, it's like water' one participant states. The facilitator probes further, to determine what they mean by 'quality'. They find they mean the actual taste of the coffee beans used. Other themes come up, such as the fact that the coffee tends to be cold when they go to get it, the machines look old, and they don't like the creamer as its powdery and gets clumpy. Also, the machines are far away, making it impossible to run to in a 5min break, and they are often in meetings back to back.

This validates the quality variable, as well as identifying what exactly it meant. The fact they mentioned that they are in back-to-back meetings, validates the segmentation data that showed employees in the office often have many meetings back-to-back and short breaks in between. It also showed the other variables that affect the experience of the coffee for the employees.

- Taste
- Temperature
- Look of machine
- Condiments - creamer
- Location of machine respective to employee desk

Ideally, you want to optimize along all of these variables, however for this example, the experience lies second to cost in the projects goal. The fact that we have these constraints and multiple variables is a key indicator that we need to see which drive is the strongest. Which variables if we improve, will increase the experience the most? This is a prime moment to do some testing.

After a bundle test (Chapter 6), the facilities team find out that taste and location are the most influential variables, because if it doesn't taste good, or they can't get to it, there is no point to it. The look of the machine is not as important as the other variables, the creamer is quite important as well as selection of sugar. The temperature surprised them. The team thought this would be very important, when in fact in relation to the others, it wasn't as important as taste and location. If they had to choose, the employees didn't mind if on occasion it was cold and they had to warm it in the microwave quickly. This allows the team to make certain trade-offs when they weigh up their options and deliver the best experience possible.

An important fact here is that the team didn't assume. They didn't assume if we change the coffee vendor employees will love it. They didn't assume if we add more coffee stations employees will be happy. This is key. If you ever hear yourself or others saying 'if we do this, then users will do this' stop, and ask yourself will they? Then test and find out. **Test your assumptions**. This leads to the next visualization.

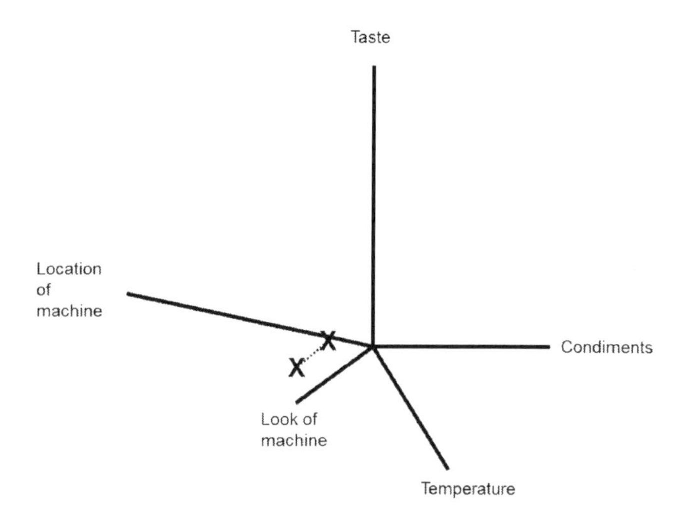

Figure 7: The different variables affecting the experience

You can see from line length what weighs more on the experience. You can also see that optimizing the machine look alone only gets you so far. This helps prioritize for example, if nothing else, that the taste must improve. It also helps determine features and a solution driven by the need. Here is a perfect moment to illustrate **knowing the needs versus listening to wants**. The team could have interpreted the user comments such as 'I would like the new machine X as the coffee stays hot for 8 hours!', by thinking that if they delivered this machine, then people would be much more satisfied. However the need of the user was that the coffee stays hot, and more importantly taste good. This could be a method to keep the coffee hot or any other machine with that functionality, not necessarily machine X - again, it is about looking at the possibilities and delivering the experience that addresses the needs.

WHO, WHAT, WHY, WHERE, WHEN

Now is a good time to think through what I will refer to as 'situations'. This is defining the different situations the solution can find itself in, so that you know it will serve the purpose as you move forward in the design.

You first list all the types of users that may be using it, what they would use it for, their reasoning, where they would and time they would.

For example in the coffee station scenario, table 5 shows some potential situations.

This helps you ensure you are testing with the correct people, it helps you think through how the solution is used and will be incredibly useful in design. From the above example alone, you can see where the coffee is located is important to the employee in the 'quick break' situation. You can see that if the stock room is far from the station, the facilities worker will need access to a trolley to carry the stock. You can also see you will need to identify how often is optimal replenish time, perhaps what are the peaks etc.

	Situation 1	Situation 2
Who	Individual Employee	Facilities worker
What	Wants coffee	Has to fill the coffee station
Why	Needs a quick break	To restock
Where	They are located in a conference room when they need the break	On rounds – stock room in building 3
When	Mid-Morning	6AM-5PM – every 2 hours

Table 5: Situations

This is an incredibly useful tool to help the thought process very early on, to help internalize for whom, and in what situations the solution will be used. I cannot emphasize enough that asking and answering these types of questions early on, leads to a much better design. It is also very quick to have a conversation or schedule a meeting to raise questions and have discussions, versus the cost and value lost in a wrong design that doesn't work for the intended users. I recommend getting the stakeholders, product owner, engineers etc. together and walk through the situations so that everyone is in alignment to the use cases and how the solution will work.

HOW WILL YOU MEASURE THE EXPERIENCE? PERCEPTION IS REALITY

This brings us nicely to metrics and measures, the backbone of business where the proof is always in the metrics. This is also very important when you are trying to influence teams or management, an important part in institutionalizing experience design. It is a dangerous subject because the numbers must be understood and often in experience they are not. Not understanding the metrics and the story they tell can easily lead to wrong decisions. So how do you measure experience in this world? Many people think measuring the user experience is as easy as having a survey question 'how would you rate the experience of X?' Unfortunately it's not quite as simple as that. You want to measure the variables affecting the experience too, this way you know exactly how it is going and where it is failing. How you measure is up to you and what makes sense for your environment and conditions. For example, if I measure the support experience I may generate a survey after the call is done. Some people call a representative selection of users and ask them, some do 'secret shoppers', some add a section to the yearly user survey… you get the point. More important for you to understand right now, is knowing what are you measuring.

There are two ways to obtain metrics: one way is to get them automatically via the system, e.g. time of a call, availability of network, or, the other way is to ask the users themselves. You could say the automatically captured metrics are 'reality', the other, the user's 'perception'. However for all intents and purposes, when it comes to experience metrics, **perception is THEIR reality**. For example, perhaps the coffee machine automatically logs the exact temperature of the coffee, and it seems the temperature is absolutely fine, yet the users are saying it's not, and they are unsatisfied. This experience perception-reality gap leads to the issue with solely automated metrics. One could also argue that the measure of satisfaction itself is wrong. Who said that the temperature measured, that the team thought was satisfactory, was actually satisfactory with the users? This brings us to another reason you need to measure the experience, because perhaps there are Service Level Agreements (SLAs) that are measured that were wrong to begin with, so they look like they are doing well, yet users are unsatisfied.

For example, the SLA of network availability is 80% off-peak, yet users are not satisfied with the network and state 'unsatisfied' in a service survey the company sends yearly. This brings us to another gap. The network team says 'but everything is fine, we are green on our scorecard', but unfortunately the perception of the users is that the network is unsatisfactory. This highlights the point nicely. The users are usually not a part of any agreements such as SLA's leading to a gap as they may not be satisfactory to begin with i.e. 80% may not be satisfactory to the users.

An experience scorecard can be created so that the area of experience can be very concrete. This often appeals to leadership, especially in companies starting out in the space. It serves to bring experience into their world, helping them internalize it further, as well as putting you well on your way to the 'do some work, prove it, do more' credo in which the experience discipline can grow within the organization.

The score card will:

1. List the experience variables.
2. List the various metrics that measure it.
3. State how they are captured and at what frequency.
4. State if they are user perception or not.

Let's continue with the coffee example to illustrate. We have the variables we found out affects the experience as listed before. The temperature can be measured automatically to ensure it is at least 70 degrees for 8 hours from brew time, the perception will be tested by some random surveying of people at the coffee station each month. Given the scope, this was feasible and valid.

Then once you start trending it, the scorecard allows you to do two things, 1) you can see where the perception matches reality, therefore finding your true user service levels for experience and 2) you can use the reality to educate as a part of Management of Change (MOC). You can make good use of the data by feeding them into any communications and marketing plans for the solution. This is also where you consider that you will need to measure these on an ongoing basis, after your launch the solution.

Now you have the scorecard you need to start driving the experience, you need to look at what you have and what you are missing. Perhaps some of these measurements are new and you will need to go ahead and baseline so that you know where you are in the space. How you baseline will be dependent on the testing or data gathering technique that is relevant to what you are designing.

You can then see where you lie in the space currently. You can also see where you should prioritize actions given other constraints, as you know which experience variables will weigh more on the overall experience. Now you know where you are, and what you need to fix. It's time to figure out how to get there.

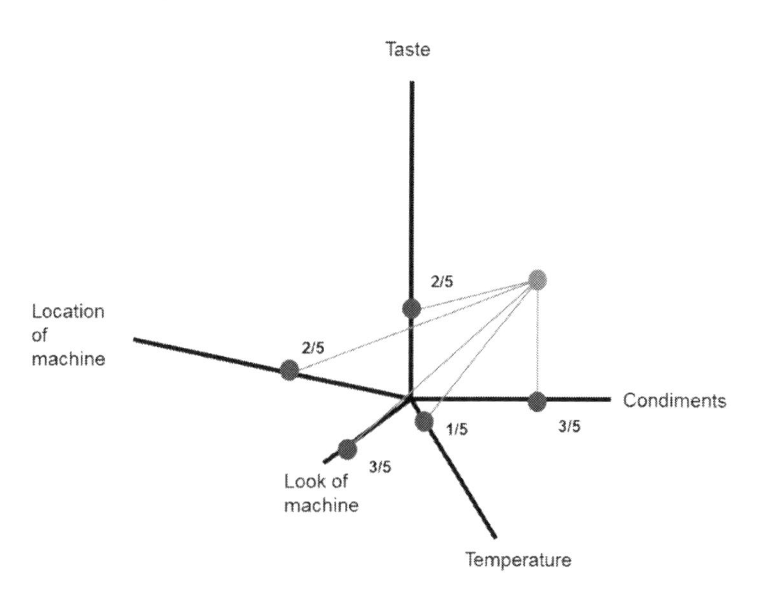

Figure 8: Experience variables and measures

Experience Variable	Metric	User Perception	How	Frequency	Baseline	Month 1
Taste	Satisfaction of taste	X	In person survey	10 people a month	2/5	
Location	Satisfaction with location	X	In person survey	10 people a month	2/5	
Temperature	Satisfaction with temperature	X	In person survey	10 people a month	1/5	
	Temperature reading	Automatic	Machine reading taken	Monthly average taken	55°	
Condiments	Satisfaction with condiment selection	X	In person survey	10 people a month	3/5	
Look	Satisfaction with machine look and feel	X	In person survey	10 people a month	3/5	

Table 6: Example scorecard

LOOKING AT THE SOLUTIONS REQUIREMENTS

Often a solution will have many requirements and you will not be able to do everything all at once. You will want to prioritize, select and design the mix of requirements that will give the best user satisfaction.

Kano Model Analysis looks at how the attributes of the solution affects the customer satisfaction and there are a lot of great online resources on how to apply it. I have found this model extremely valuable in enabling prioritization on the correct features to optimize the solution, helping stakeholders make the best decisions, keeping teams focused throughout design, discovering solution differentiators and identifying customer's needs.

The Kano Model plots the attributes over two dimensions – the amount of investment (how much the attribute is implemented) and customer satisfaction. The model divides the attributes into categories so that you can understand which will wow users and which they will expect.

Attribute Categories

The categories are defined in the following ways:

- Attractive, wow or excitement attributes: the customer will be very pleased with these attributes, however if they are not there it will not cause them to be unhappy. You need to balance the amount of 'wow' types of attributes, as without the basic attributes being present in the solution, the user will be unhappy anyway.

- One-dimensional or performance attributes: the more the attribute is implemented, the more satisfied the customer will be.
- Basic or must-be attributes: attributes that the customer expects to be there. If it is not there they will be dissatisfied, however, if it is there they will not be more satisfied. These are 'musts' for the solution, but are likely not differentiators.
- Indifferent attributes: the customer will not be any more or less satisfied if this attribute is present.
- Reversal attributes: customers will be dissatisfied when this attribute is fulfilled and satisfied when it is not.

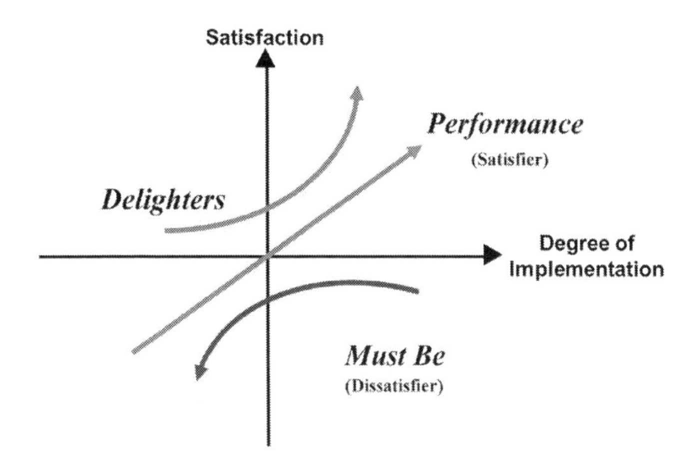

Figure 9 : Kano Model

It is worth keeping in mind that over time attributes that once delighted or wowed a customer may migrate to the expected category.

Applying the Kano Model

To apply the model there are two questions for every attribute that you will need to ask your customers:

1. Functional question: How they would feel if the solution had the attribute?
2. Dysfunctional question: How they would feel if the solution did not have the attribute?

Customers can only respond with one of five responses:

1. I like it
2. I expect it
3. I am neutral
4. I can tolerate it
5. I dislike it

You then analyze the responses for each attribute and each user questioned. A handy table for your reference is shown in table 7.

		Dysfunctional Question				
		Like	Expect	Neutral	Live With	Dislike
Functional Question	Like	Questionable	Attractive	Attractive	Attractive	One-Dimensional
	Expect	Reverse	Indifferent	Indifferent	Indifferent	Must-Have
	Neutral	Reverse	Indifferent	Indifferent	Indifferent	Must-Have
	Live With	Reverse	Indifferent	Indifferent	Indifferent	Must-Have
	Dislike	Reverse	Reverse	Reverse	Reverse	Questionable

Table 7: Mapping the responses to the attribute categories

After, you can categorize the attributes and work with the stakeholders (product, engineering, marketing etc.) to ensure prioritization and designs are in alignment with making the solution successful.

PUTTING IT ALL TOGETHER

Changing an experience or creating a new one can take some time, especially when budgets are given to competing goals which can be very common. It is good practice to think of the steps it will take to get there. Then, given constraints, think what can you achieve given you know what affects the experience most, maximizing your potential gain in experience. To recap, we now have most of the blocks to feed into the planning stage.

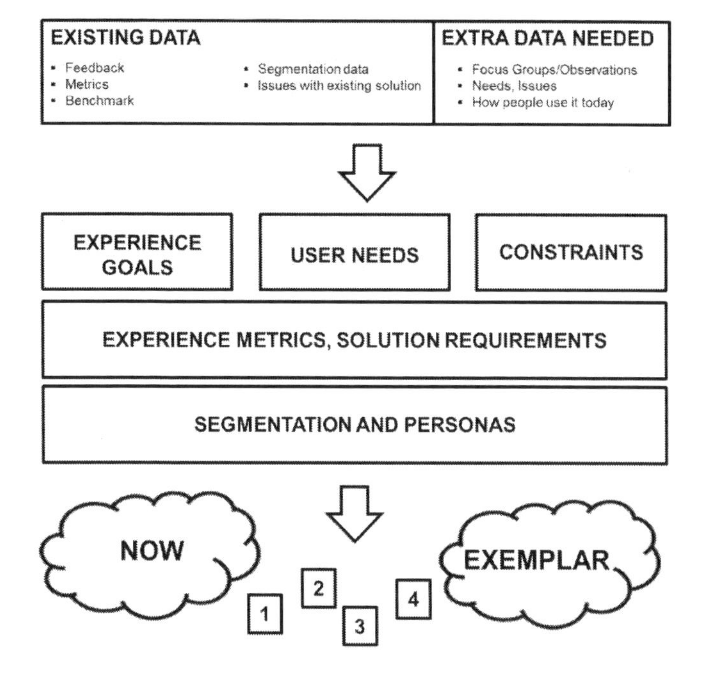

Figure 10 : Experience planning blocks

We just saw how to obtain the baseline. This tells you where you are now, but, how can you plan to get somewhere unless you know where you are trying to end up? The next step in the planning is logical - where are you going?

You have almost everything you need to know to define this. You have all your input data, you have the need and the goal. You now need to think though and validate what is exemplar? What is the best experience in the end user's eyes? What would it look like if those variables were all 5/5?

The important word here is 'validate'. **Again do not assume**. You may have the information already from previous testing in the data input stage, but if not, test for what exemplar is. Conduct some scenario testing or concept testing with your end users to see what exemplar truly is. Taking a moment very quickly to discuss the concepts with your end users is extremely valuable.

Now you know where you are and where you are going, you can see the gap and start defining the steps to get there.

For example, perhaps the exemplar coffee station experience could be described for the different user types as:

Employee: Hot, tasty coffee within 5 minutes of my location with full range of condiments (sugar, creamer).

Maintenance: Coffee machine that alerts when product supply is low, with inbuilt clean cycle function, near the supply room.

Now, think through what will have to change from your current state of 'now' to get to the 'exemplar' state. This is where UX can help drive the solution requirements and specifications. You can add user requirements to the solutions requirements document and help prioritize using the Kano Model previously discussed.

To get there, the facilities team will have to:

- Supply tasty coffee.
- Ensure coffee stations are conveniently located throughout the office.
- Ensure machines have a clean and alert functionality.
- Ensure the coffee is hot, etc...

Now is when you start road mapping. I realize we are using a trivial example, however it demonstrates the thought process perfectly. So for now, we shall create the experience roadmap for the company's coffee stations.

Given the team's constraints such as budget, they need to know what the employees would value sooner, and what, if they had to, they would wait for and still have a noticeably better experience than today. They need to look at all of the feasible options that would improve the experience for each requirement. For example:

- Tasty coffee: Starbucks beans ($$$), market generic ($$) and coffee x ($)
- Temperature: Manually reset temperature (-), new temperature controlled machines ($$$)

Again, this isn't any old option, these are feasible options given they know what they have to improve and what is important. For example, perhaps an option is to use Starbucks beans, but then in phase one of the project they wouldn't be able to do anything else due to the cost. This is a great time to do a bundle test with a timeline and cost (Chapter 6). The facilities team conducts a bundle test and their steps for them to execute, given funding and timeline, are:

Phase 1	Improve coffee taste to market generic, no machine replacement, plus ensure temperature gauge is functional on all machines, change heat pad to reheat at 69°.	2 weeks
Phase 2	Add range of condiments to include creamer pods and Equal, Sweet and Low and sugar.	1 month
Phase 3	Add more convenient locations - 2 more per floor.	6 months
Phase 4	Machine replacement role out to new machines.	2 years total refresh

Table 8: The roadmap

And there is their roadmap.

4

ORGANIZATIONAL ALIGNMENT AND STRUCTURE

ORGANIZATION SET UP

For those that are just starting the journey to institutionalize user experience in their organization, there will be a time when you need to consider the organizational structure as a factor in success, as well as what ideally you would like to grow it to. The level of organizational maturity in user experience will play a large role in how you implement the framework, and to what extent. If you try and create the ideal world straight off with all the trimmings, sometimes it's seen as too complex to management, so start small and get to quick provable wins. That being said, you yourself should be aware of how you will grow UX principals and work effort throughout the organization, or at least have some vision as to where you want to take this discipline, almost your own roadmap. For example, when I started I knew I wanted to spend a few months proving the value in ad hoc, highly visible projects. Then I wanted to start a team, then a process to spread it to other groups. This was my plan, knowing how mature the organization was when I started. In some cases maybe you have the team already and it's about process, or in some cases you can do it all at once. Sometimes, for example in small companies, you will be the UX person and play the role across all facets. In this case you would want to know where you play your part and who you will work with to achieve the UX vision. Also worth thinking about is the types of work and skills needed in a UX team (if a team is in scope for you). You would want to consider people for user research activities, usability testing and design work. Each brings various skill sets and knowledge of various methodologies and tools.

One of the biggest reasons for failures in experience initiatives is a lack of organizational alignment. This is especially true when you need other teams to commit to work, resources, or improvements with pieces they may own that affect the experience for which you are designing. Sometimes the span of direct control is not always within your own team, or even department. For example, for HR to create a new employee start up system, perhaps they need help from IT to improve the system that they own as a part of the startup for a new employee. The truth is, there will always be competing goals, this is a reality, and sometimes experience goals can be put on a back burner. This takes us back to one of the fundamentals we spoke about in Chapter 3, **don't lie to yourself as to where the creation of a great experience lies within the goals**, and always aim to create the best given the constraints. This not only allows for achievement, but makes your plans seems far more achievable to other teams and management in order to align them.

The first question to ask yourself is what would it look like ideally within your organization, if experience was ingrained and just simply the way you worked? There are two good set ups I've seen that allow for experience to be either with some level of tops down 'ordinance' or at least fully-fledged so that it has a good seat at the decision making table, as well as some strength in the competing goals scenario.

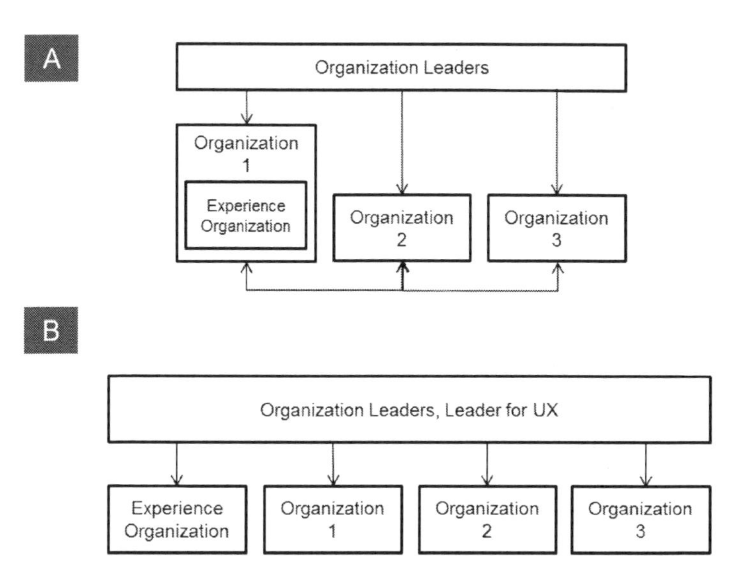

Figure 11: UX team interactions

In Image A, experience is embedded inside another organization, with an overall leader e.g. inside Engineering. UX would still work horizontally with the other organizations. In Image B, there would be an overall UX leader, for the UX organization which is of equal level as the other organizations. The truth is, it is unlikely to suddenly occur this way especially in larger organizations or those that are less mature in their level of UX.

Therefore you may well have to grow it 'bottoms up' to a point where you have proved it works.

Two typical set ups I have seen to take this approach are 1) to embed an experience person within a project team for a specific project that requires it or, 2) to have a central team of experience experts that are assigned to different projects within an organization.

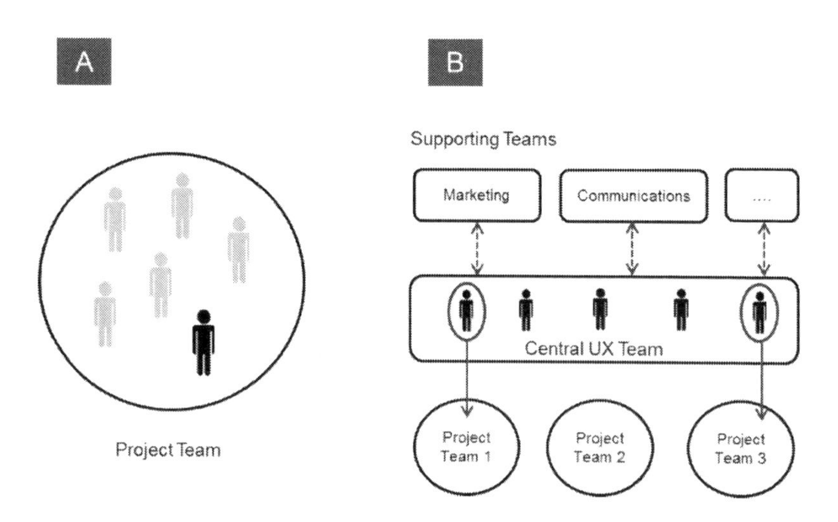

Figure 12: UX resources in projects

This set up can be replicated in each organization that needs UX, for example in different product teams. Then you can see how horizontally all people working on experience are aligned without a central team. Frequent conversation between these individuals can help keep UX in alignment.

This is often the more achievable placement for experience experts; however, with continuous achievement and proof of it, it is normal to grow into having an overall experience organization. Either way, remember there will have to be strong linkages into the other teams that affect the experience e.g. product, engineering, communications, marketing etc. Even better, consider a person within these organizations that is accountable for experience, and the working relationship with the UX team and individuals.

If you are to be utilized in specific projects, the best way forward is to select high profile projects and concentrate on a measurable impact to experience. You may play several roles, especially in a lean environment. For example, you may do design, run the usability tests etc., as well as engaging and aligning with the other teams involved. If you are in a centralized team, then select and prioritize projects depending on impact, as likely in a lean team when starting out, you will not have capacity to do every single project that requests your resources. Therefore, you will want to think through your criteria for prioritizing. In table 8, you may focus on Project C as it is high impact, highly used and sales increases the company's profit. You may also choose Project B, as it will be used by leaders quite often and you do not want a bad experience at their level. How you choose will be based on what is important to your company at the time.

Another great piece for a centralized team to own is collection of user data and research, be that their own or collecting materials from other departments or from other companies. They should provide continuous study of the target users and their behavior, needs and profiles, all of which can be utilized throughout the ROAD™ framework as an input data.

	Project A	Project B	Project C
Number Users Impacted	50,000	5,000	200
Impacted Segments	Corporate Functions	Senior Leaders	Sales
Frequency of User Interaction	Once a year	Couple times a month	Every day
Amount it Affects Overall User Satisfaction	Medium	High	High

Table 9 : Example criteria for project prioritization.

Types of activities a User Research Team can conduct are:

- Maintain data on users and their demographics.
- Trends in user behavior e.g. mobile use, how they interact with technologies, emerging innovations.
- Market trends.
- Current research in related topics and benchmarks.

- Carry out study's on the user base such as segmentation and day-in-life.
- Create Personas.

Sometimes HR organizations or external research organizations have a lot of great data on users, so perhaps setting up linkages into this is a good place to start, or subscribing to some of the major reports.

HOW TO INFLUENCE

A key part of the job is influencing all involved, and this includes alignment and ensuring everyone is on the same page to create the wonderful end to end experience for whatever you are designing.

For this, the first step is identifying who 'everyone' is. You need to identify all the key players. That means owners for marketing, communications, leaders or management in the teams, project team members etc., basically whoever is needed for the project that will need to be on board and supporting the experience efforts.

You will want to have a meeting to start the project with them in which you will want to 1) align them 2) influence them so that they buy into the importance of, and how to conduct experience design. This is your moment to show the value. DO NOT BE FLUFFY or sound like you are a psycho-babbling loony. So many times I have seen initiatives fail because the experience expert cannot influence leaders and project teams to understand the tangible benefit of this work. Instead, they quote off of slides that talk about the standard theory such as saving money in development etc. Before you go in, know who you are speaking to and what will make them specifically see the value. For example, if you are speaking to a product team, perhaps go in with an experience evaluation of their current product from their users. Then, show them a mock-up of what you see as a better experience, perhaps even with some concept testing completed. Using these measures are far more powerful to demonstrate the importance of user experience and thought in design as a differentiator.

During the meeting, items that will need to be discussed and agreed upon include:

- Experience goals.
- User needs and values.
- Metrics - current and goal.
- Where experience exists within the other goals.
- Features and requirements, in reference to user need and impact to end users e.g. show them the scale of experience and what the impact will be (such as, if you are going for the cheap coffee this means x on the impact, if you miss the pause feature on a new headset, then this means y).

You will also want to go through the process with them that you are going to use, e.g. strategy, design, delivery and ongoing phases, as well as how you will capture data etc. Keeping it high level at this point is very important, they don't need to know the details of everything, this can come across as a little too 'fluffy', especially in less mature organizations.

You want to ensure you go through the dependencies with other people and projects e.g. the communications lead will want to know their part in this. This is to ensure everyone knows their role and are in agreement. You may want to consider the use of a RACI chart. RACI's show who is *R*esponsible for work, who is overall *A*ccountable, who is *C*onsulted and who is *I*nformed.

	Project Manager	**UX Team**	**Development Team**	**Communications Team**
Measure Current Experience	A	R	I	I
Design New UI	A	R	I	I
Develop the New Solution	A	C	R	I
Create Communications	A	C	I	R

Table 10: Example RACI chart

In RACI charts you should only have one A in the row as only one person should be overall accountable.

You should facilitate this meeting, in which you will often catch misalignments and competing goals where one teams interests are not another's. Once decisions are made, you should obtain sign off from all key players.

ENSURING VISIBILITY WITHIN PROJECTS

Like I said, sometimes, especially in big projects and especially if you are a resource embedded in a project, other goals take priority, or the experience work may quite frankly not be taken seriously - especially in less mature organizations. A good way I have found to deal with this is to ensure all dependencies and milestones for the experience work are on the Project Managers overall project plan. This ensures it is visible and brought to the project manager's attention, it will also help you see dependencies on other teams and people. Some examples of milestones to add are:

- User flows complete
- Usability testing complete
- Roll out plan complete
- Etc...

FINAL COMMENT ON INSTITUTIONALIZING EXPERIENCE

Remember, when you start, do some work, prove it works and repeat until you grow it through the organization. Present on the process and benefits (with proof), to teams that you think are instrumental in experience and that may want to start focusing on it. This strengthens the network and need for it, which eventually will capture management attention.

And finally, get a sponsor for the work, an executive or high level manager is best. Find someone with some clout in the organization, someone who is a believer in experience work, understands its importance and value who will advocate it, and help you institutionalize it.

The rest, from what I have seen, is really up to the person spearheading this work. Some key learning's I myself have realized along the way are:

1. Know what you are talking about better than your audience (this book will help you with the main fundamentals, but you will need to continue learning by using all the other great resources and books that exist). For example, if you are talking to a mobile application team, know the latest trends, the experience of them, what is coming next, etc.

2. Do not send a boring, monotonous person to present on your process. Engaging and exciting your audience about UX and its value are keys to gaining their buy in. I know many 'snooze worthy' professionals who are amazing at the work they do, but are not for putting in front of an audience to influence them. UX is amazing, and you need others to feel that too.

5

USER FLOWS

These are one of the most powerful tools in designing any solution for your end users, especially in organizations with multiple teams contributing to the solution creation, or where there are cross team dependencies.

Essentially user flows allow you to map out the user's path through the solution, to vision their end-to-end experience and see where they can go or end up. They also allow easy and quick detection of gaps and questions to be raised early enough to save valuable implementation time. The truth is, in many organizations once a solution is implemented it is incredibly difficult to make changes as perhaps resources have moved on to other projects, or another iteration has to be waited for, therefore, user flows are even more valuable. **Why do something wrong and fix it later, if you can make it as good as it can be now?**

It is worth mentioning that you may hear all sorts of names and terms to describe the action of describing how the users interacts with the solution. User journeys are a term you will hear a lot. They are essentially looking at the steps of the user's interaction with the solution including context, devices, functionality and emotion. I started using the word 'user flow' to encompass it all simply as I found it appealed and resonated better with engineers and leadership.

It is valid for almost everything – take a soft drink for example and a user that is in the 'about to purchase' flow. Their flow could include 1) noticing the product 2) picking up the product 3) reading the label (nutritional information etc.) 4) checking the price and success would be 'purchasing the item'. When you understand their choice points, moments they could be in error and how to get them to success – you will get them there.

WHEN?

When would you do these flows in the creative process you may wonder? Well the easy answer is as early as possible as starting before any initial building or implementation will get you the most value. Normally there may be an investigation or analysis phase, that creation of these flows would be perfect for. The other key point to remember is that these flows are living documents that are likely to be added to and edited as decisions are made and the solution changes and grows. Even goals can change mid-way, or extra features added, so remember to maintain the flows throughout the creation process or life cycle.

GETTING STARTED

The first important fundamental to understand is that user flows can be done at varying levels of granularity, they can also be pieced together e.g. one user flow can point to another.

Sticking with the example of coffee (which is like a programmer's 'hello world' to this book), on a large scale the experience of 'getting coffee' can be described as all major steps to obtaining a cup of coffee. This includes 1) getting to the coffee room and 2) making the coffee. On a more granular scale, the experience of actually 'making the coffee' can be broken down into sub-steps, and those steps broken down again, and so on and so forth until the whole process is laid out.

The key here, is to start with the end users goal which will be their successful completion of the task with the solution, then break down each step the end user must do to get there.

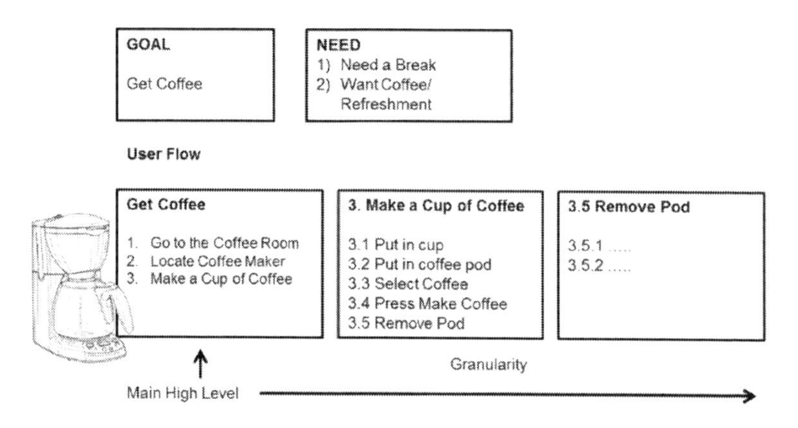

Figure 12: Break down of a user getting coffee

Let's illustrate the coffee example a little.

Figure 11, depicts what is normally the successful flow - e.g. what steps the user goes through when everything works out perfectly as planned and no erroneous situations occur. As ideal as that is, there will be times when things can go wrong. To ensure a great experience, what the user sees and experiences in these situations are vital.

ERROR FLOWS

How the error states are dealt with can sometimes be the biggest factor in the user's experience. Flows are a great way to ensure you have thought about any potential error situations that a user may find themselves in, and what will happen to them.

If, for example, something is broken, how will the user know and how will they move on past the error? Perhaps there is another action for them to take such as call support, or perhaps it is an easy fix that they can do for themselves and get back onto the successful path, in which case how will they know what to do?

A good way to think through this, is to look at every piece in the flow, and ask yourself 1) is there anything that can go wrong here that will stop this step from occurring 2) if so, how will the user know what happened and 3) what is the next step for them.

For example, in the coffee illustration, perhaps the coffee machine needs a clean cycle and therefore when the user presses the 'make coffee' button, nothing happens as the clean cycle must happen first. In this scenario, perhaps a message is displayed to the user to press the clean button.

It is easy to see how this is translatable to other solutions such as a website or system. For example, if a program is copying data for the user, and for some reason the connection is lost, what message does the user see and what do they do next? The same questions at each stage apply.

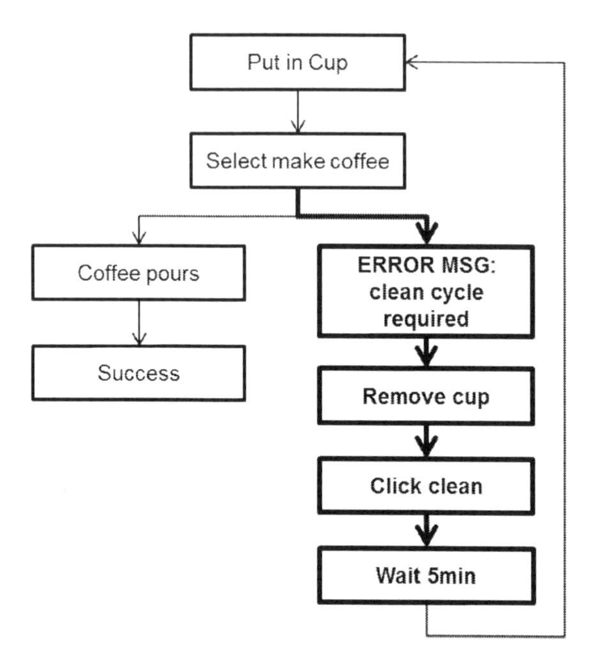

Figure 13: Error flows

PRE-REQUISITES

These are things that must be true for the user to even be able to enter into a stage of the flow. For example, if the coffee machine is broken they cannot use it, or if there are no coffee pods, then they cannot make coffee.

Here, you must think through each in the same way as an error - what does the user see and what can they do about it? Sometimes pre-requisites themselves may be a flow of their own as per the example in figure 13 where 'refill coffee pods' would occur if the prerequisite was false.

3 Make coffee	**5 Refill Coffee Pods**
Pre-req 0.1: machine working = true else go to 7: Submit Support Request ***Pre-req 0.2: coffee pods available = true else go to*** ***5: Refill Coffee Pods*** 3.1 put in cup 3.2 put in coffee pod 3.3 ...	5.1: get pods from draw 5.2: place pods in pod tray 5.3: complete pod stock form

Figure 14: Pre-requisites and joining flows

CHOICE FLOWS

It is also important to add in any choice paths e.g. where the user has options that may affect their path given the selection they make. Also you must consider what information they may need to select the right option for their goal, what time they need it and the best medium for them to have the information delivered.

For example, when the user inserts the coffee pod into the machine they have a choice to make - they must decide the flavor of the syrup they want to add. This choice must be made before continuing. It is a definite decision point for the user, however it is non-course altering. How the user continues on the path after the selection is made does not alter, only the outcome is altered e.g. what coffee they get.

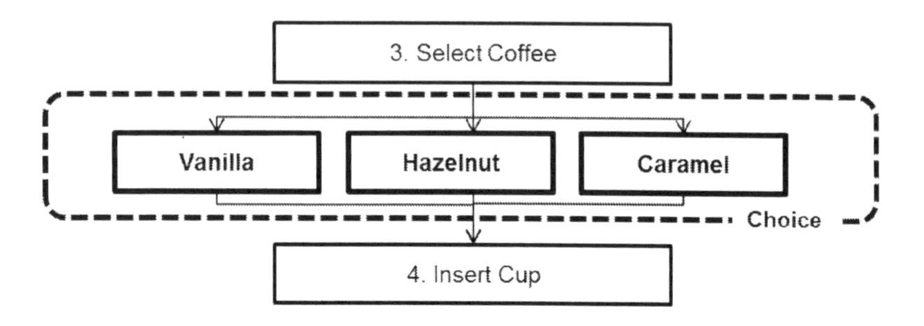

Figure 15: Non course altering choice flows

Normally speaking we do not want to overload the user with too many decision points, as too many options can be overwhelming, especially when they have a simple task to complete. An example I love to give is, it's almost like knowing you want a turkey sandwich on wheat and then being asked what bread you want (wheat, white, rye, grain, 5 grain, sourdough), what cheese you want (cheddar, American, Swiss), what toppings you want, what sauce you want... When all you want is a turkey sandwich and fast... You get the point. The key is having the right options (you can test if they are correct in your concept tests) and, having a quick way to select the options. For example you may have seen the paper selections in Cafes where you check mark the box for what you want. You have to consider both scenarios – the user has no clue what they want, so advertisements and hearing options are good for them or, the user does know and therefore just needs to get it fast. This is like thinking through how a first time user flows through vs. a frequent user. Where the sandwich shop goes wrong is when they train their employees to ask all the questions, always, to everyone. Another great example of the **importance of considering your user, their task and flow through the process**.

Remember, sometimes decisions users make may alter their path all together, in which case you would want to map the path from the decision point. For example, in the HR system for careers, which has the goal of enabling users to get help for the next step in their career path, the path they take will be dependent on their need. This is because depending on the person's time in life they could know what they want and want to search for a job, or they could not know what they want and therefore want to read about different options. Translating this to the tool, on the entry page if a user selects 'help me figure out my career' they may go to a wizard, and if they select 'find a job', they may go to a search tool. They will be successful in their goal at a finite stage in another flow depending on their choice. This is demonstrated at a very high level below. The user who wants to explore will follow the explore flow and can exit at that point (of course you can have a return to 2 to view options).

Goal: Find Next Job	
1. Log onto careers@companyx.com 2. View options 3. Select option	

If 'job wizard' then flow 3.1 If 'explore jobs' then flow 3.2 If 'job searcher' then continue 4 4. Enter search criteria 5. See results 6. Look through jobs 7. Apply Exit	3.2. Explore jobs 3.2.1 Select criteria 3.2.2 Read testimonials 3.2.3 Exit

Figure 16: Choice flows

JOINING FLOWS

For many experiences and solutions flows are too big to make one big all-encompassing flow, therefore flows can be linked. This also helps if different teams in the organization are responsible for different parts of the flow. One thing to remember is to have some governance on the tool used and naming conventions, this way the flows can be easily pieced together and combined for discussion, analysis and testing purposes.

LINKING IN TRIGGERS

At the beginning of the flow it is useful to think through the different entry points, how the user actually comes to enter the flow in the first place. For example, were they sent there from somewhere else like a link on a site, did they find it through some need, or was it a trigger of their own like their desire for a cup of coffee. This could affect their path, so it is very important to consider.

Here is another example: If a user is going through the process of ordering a new phone, this could have been triggered because he is due a new one as his contract ended, he is a new employee and is ordering needed equipment, or his existing phone is broken beyond repair. All of these will affect his path as an end of contract could be triggered by a renewal notice, and also means the user will still have his old phone to get rid of so that may be another flow that he has to go through after ordering, and a broken phone means he may have to file a report before ordering a new phone which will alter his path.

You can see now how a concept so simple, actually is very valuable indeed to raise questions and close gaps in the experience. Just remember, where is the user in the flow, what can the user do and where can the user go?

COMMUNICATIONS AND MESSAGING

Another great use of flows is seeing where communications and messaging are needed within the flow. What information does the user need and which medium should be used to relay the information? After all, you can have the greatest solution, but **if people do not know it is there then what is the point?**

Let's look at an example for the coffee machine. The machines are new, and the facilities team have done all this work to create a great coffee making experience for the employees; however, how will the users know? How will regular coffee makers be aware of the change and not be surprised on their next visit that what they were used to doing and could do easily and quickly has changed? How will users that didn't use the coffee machine know of the changes in case it can entice them to use it now? These are things that will be considered in the delivery plan - which the flows will be input to (Chapter 7 Delivery). When messaging and communications are shown to the user these should be at least highlighted in the flow.

Looking back at the error situation of the coffee machines clean cycle, you must consider that the user may not be a savvy coffee machine user, and may not realize what will happen. Perhaps this leads to a poster on the wall right by the coffee machine to explain what the clean cycle is. An instructional poster that tells the user 'If you see this then this is what it means, and this is what you do'.

WHEN ARE FLOWS COMPLETE?

Flows are complete once all paths have been considered and have a finite outcome, even if the outcome is not the successful state. For example, if the search is broken on the career tool due to a network outage, perhaps the outcome is to simply call a support service. While the flow is not in a successful state, the flow is in a finite state, and you know you will need to inform the users of their steps to contact support at that time.

Linked to this is the fact that a finite end state must be stated e.g. user has cup of coffee. If given errors or choices and the path leads to other flows, then if you follow the next flow that was linked to, the user should still come to a finite state.

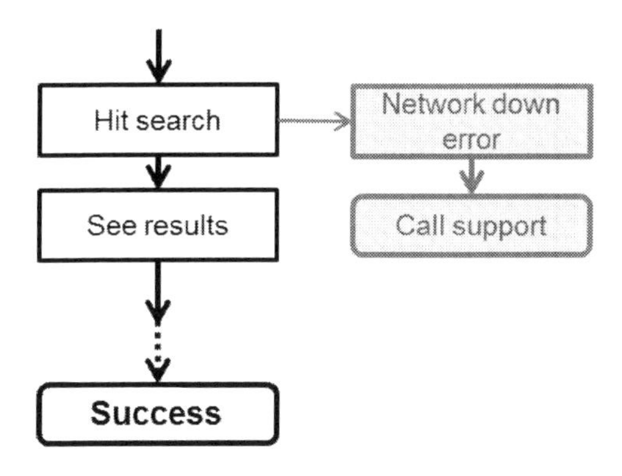

Figure 17 : Finite states in flows

The power of flows are in limiting user confusion, understanding user choice points and the potential error flows. When working with flows there are four things you should ask yourself at each stage:

- Where am I?
- Where have I been?
- Where can I go?
- Which is the best next move (and how do I know)?

These play a huge role in guiding the user though the path to success. It enables the use of context to make the path easier and helps the user understand his choice points and avoid potential confusions and/or obstacles.

COGNITIVE BARRIERS

A cognitive barrier is what stops an end user from getting to where the user wants to be. There are three cases that can affect the ease in which the end user travels through the flow. You can use the flows to look at balancing these cases, which are:

- Number of steps a user has to take.
- The user's perceived length of each step.
- The user's perceived difficulty of each step.

Most barriers are temporary, and can be overcome through some 'information processing' by the user. For example, a step can be difficult, but there is step-by-step guidance explaining how to do the step so the user doesn't have to be confused. If the user cannot overcome the barrier, and cannot figure it out, then there is a higher likelihood he will abandon the task he was trying to achieve. This has a profound effect as it can damage the solution's reputation through the user's perception, and decrease adoption thus decreasing the proposed value.

Number of Steps

It is not necessarily true that 'the fewer steps there are, the better'. It is very important to know when steps should be added. For example, if you are installing something on a user's PC and they click install, then nothing happens as it is installing quietly to minimize steps, the user may wonder what's going on. In this case adding a progress bar and an 'install complete' notification with an 'ok' acknowledgement step may be better to indicate success.

Length of Steps

There are two things that are important in balancing the length of steps. Again it is not true that the shorter the step the better. You need to look at the user expectation of the length of the step and the cognitive load the step places on the user. Users will spend the time interacting with a solution that they understand, and you need to understand when a user is ok with spending longer on a step. For example, in a system where the user is storing personal data, the step to 'sign up' maybe take 5 minutes while he thinks of security questions, as opposed to just a simple user name and password that takes 2 minutes. A user may be ok with this as the step is perceived to be important and for a reason the user can understand. You can use your flows for testing with end users to see where steps may be too long, or where it is ok for the user to spend some time on a step.

Difficulty of Step

This is quite subjective as different users will find different steps more or less difficult depending on their experience and knowledge. Again steps can be difficult, but made easy to complete via guidance, and if you create the step for a spectrum of users, then you can make the guidance a choice - that way the user who is more knowledgeable can go on, and the user that needs help can have it. Users are more likely to persevere and complete a step if they understand why it is so difficult.

COGNITIVE LOAD

Cognitive load, is the 'amount a user has to think' to complete a step. Things to consider in balancing the cognitive load are:

- The number of choices at a given step.
- The amount of thought required.

Number of Choices

Decision points are about helping users make the correct decision. What do they need to know, and what can you tell them at that point with what you already know of them. The human working memory is limited, and if there is too much choice then it is easy to get overwhelmed and give up. It is also easy if not guided to the correct choice to try one, go down a wrong path, then go back and try another. This cycle will lead to frustration and a higher chance of abandonment.

The amount of thought a user has to put into a step is important, and varies in humans. One user may have to think more than another to complete the same task. It is important here that sometimes you need to make the user stop and think, and that's ok. Users will be ok with that if they understand why and the importance of it. For example, ‘so that people cannot steal your bank information, please take a moment to write 3 security questions and answers’.

Confusion and choice play a large role in ensuring whether a user can or cannot get through the flow. They are likely to abandon a flow if they do not understand their options and they are likely to be frustrated if they perceive the step to be longer or harder than they expect given what they understand of it.

As you walk through the flows with users, they can help you identify where steps are longer than they need to be, where they need more information etc.

ILLUSTRATIONS

We have been using example flow charts and boxes to show the flows. Taking the flows a step further each step can be illustrated and annotated to really illustrate the user's experience. If it is a piece of software or has some UI these can be included. If it is a physical service, images or sketches can be used. A trivial example is shown below – but you can see how using illustrations or sketches allow you to really visualize the experience your users will have, where they will feel various sentiments, potential error etc. You can use these as storyboards to show them the experience and get rapid feedback on the flow and steps.

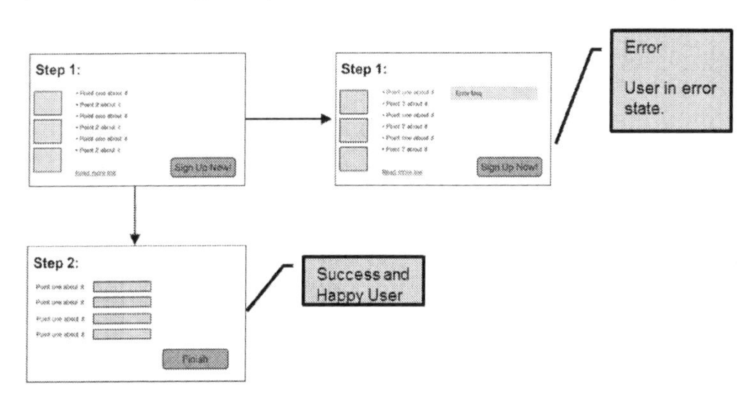

Figure 18: Conceptual example of illustrated user flow

DEVICE LIFECYCLE AND INTEGRATIONS

The increase in devices and the connected world we live in means that what device users are using to do which tasks must be taken into consideration. With the way in which data and actions now traverse seamlessly between applications, how the solution integrates and interacts with other solutions is also key. It is worth mapping these points out in your user flows.

For example the device lifecycle considerations for a user completing the action of 'booking a hotel room' could include:

- A mobile device to search and browse for the best hotel offer, as they can do this easily on the move.
- A laptop to set up a user profile and make purchases, as this requires more input and they perceive it as better for financial transactions.
- The application could integrate with another site so that the user can book the hotel straight from the review site without having to go to another application.

NEXT STEPS FOR FLOWS

You will use these to bring the various players together to answer questions, come to agreements and close gaps. If you yourself are not the person implementing the solution, you will want to likely have meetings as you create the flows with people who can help you answer questions such as what happens here, what are the options etc.? When the flows are complete, it is a good idea to get all the players (stakeholders) in a meeting and walk through the flow to show them the users' journey. This helps alignment and ensuring everyone is on the same page with the design of how the solution will work for the end user, especially when working in cross-team environments.

The flows will also be input to the communications plan as well as the testing plan - as you will want to use the flows to ensure you have adequately covered all of the use cases by selecting test cases that go through the different paths in the solution.

TAKING IT A STEP FURTHER...

The whole point in creating things for users, especially when it comes to systems and tools, is to **make the user faster and smarter in the completion of the task at hand.**

A great way to use the flows to help see ways to incorporate this consideration into the solutions design, is to go through and see what automation can be done, what work the system can do for the user and what can lighten the load, or even wow them.

For example, if a coffee machine needs fixing, perhaps maintenance get an email which stops them from manually checking every hour, or relying on a user to submit the information. This makes them more efficient and takes out work effort for them. Other common examples are form data entry such as pre-populating fields depending on who is accessing it, or based on user choices giving recommendations. Perhaps if a user is on a mobile network, then context such as location can be used to give useful information to the user.

The way to think through this is at each stage; ask yourself, what does the solution know of the user? What can it do for the user at this stage? **How can you use known context to enable a better, adaptive experience depending on who the user is, their environment and what they are trying to do?**

Today, we are even more enabled to create intelligent experiences due to the vast amount of data we can access about users. Sensors are getting cheaper and cheaper and data such as location, proximity, previous actions etc. are all readily available. It is worth it to spend time thinking through the information you have or could have that could make the task at hand even easier.

To bring all of this together – you can think of the task I like to call **'Flow Optimization'**. This is the process of looking at each step in the flow and balancing time, choice points, automation opportunities, context, information available and optimizing all of this to create the best path through the solution for the different users and use cases.

6
USER INPUT USABILITY AND OTHER USER TESTING

One of the most memorable moments I had was hearing a great, true story that encompassed the core mentality of experience design within an organization. The designer of a product asked a VP of the department responsible for the product 'what do you think of this, is it right?' They were looking for some form of validation that it was correct and ready to move to production. The VP replied 'I don't know. Why are you asking me? Go and ask the people that will use it.'

And that, is the essence of user testing. It may seem like common sense, duh, test with the people that will be using the solution you are designing...however it is a very common occurrence, especially in less mature experience organizations that a higher-up manager knows best, or one person or one team does. The truth is, while we can all be considered a 'user', **we are ultimately clouded by our own subset of experiences** and interpret all design through our own schema (the way we view the world). For example, I am a shopper and I know my own pains with shopping experiences, but it doesn't mean I can design a new successful shopping experience on that knowledge alone. Therefore, usability and other user testing is needed to allow us to extract ourselves and our experiences from the design, allowing us to look at our intended user population - their behavior, what they do and how they think. It is worth a mention here, that solutions must function. Functionality and getting something to actually work is step 1. This sounds silly, but I cannot count the vast number of times solutions are not functional, let alone have a great UX. Think of functioning features as a key usability requirement, for example loading fast enough. Otherwise these 'functional' issues will cloud your user tests.

TYPES OF TESTING

There are several times when designing a solution that you will need input from your user base to know that you are designing the solution correctly for their needs. Notice how I said needs, not wants – a key concept, which cannot be emphasized enough.

Some of the most useful types, especially in a trimmed or lightweight process are:

- Testing the what: Concept testing (storyboard), perception testing (brand value).
- Testing the how: Usability and design (actual product).
- Testing the importance and priorities: Bundle testing.

CONCEPT TESTING

Concept testing happens when the solution is an idea, likely in the roadmap stage when you are figuring out the 'what'. The goal of this testing is to see if the idea is right, if it is even meeting the user needs or solving their issue. It can also be used to test different solution design options.

Example - coffee break room

The need: Get coffee, have impromptu conversation and/or take a quick break.

Your concept: You are thinking of a room with 3 round tables each with 4 seats so that people can take their quick coffee break and also converse if they so wish.

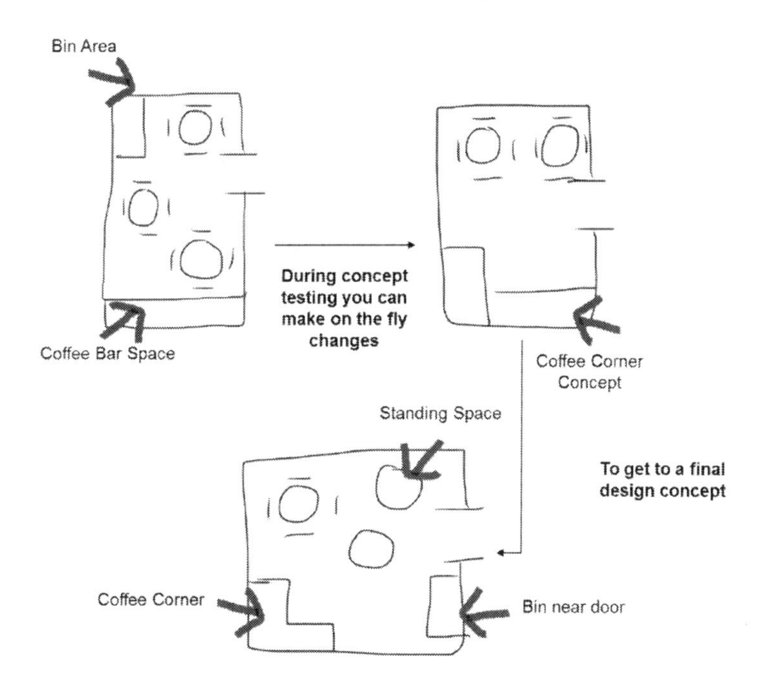

Figure 19 : Concept testing

You do some concept testing and see that no one would use your seating; in fact, they think it is a nuisance as it's an obstacle to the coffee area. Also, they wouldn't necessarily need seats in the room, as their need is more to stretch as they break, and have a fast conversation if they need to converse at all. You move it to the side, now seating is out of the way and there is still the option to sit for a longer break.

Concept testing can take a variety of forms - a storyboard, or even an open discussion, but the main components are the same. You explain, mimic or illustrate the concept and see what the users think, what they like and don't like, if it addresses their needs and more importantly, finding out what would they do with it. It is a similar concept to the **fail fast and often, cheaply**. You can use drawings that you can change on the spot, so there is no need to even have a working prototype. You want to keep this conversation very open, so that you can really learn what they will do and not just what they say.

This is a great chance to also test the assumptions that you made when coming up with the idea or concept in the first place. Sometimes people do not think they made any assumptions. An easy way to detect some of the assumptions that you probably made subconsciously or without thought, is to talk through your idea and see for the times when you describe your design that you say things like 'if this then that' or 'users will'. For example 'if we have round tables then the users will sit there for their coffee break'... but will they? This is an assumption, which needs to be tested.

If you do not test that your assumptions are valid, you can get unintended results or limited usage of the solution as people may not adopt it. This can cause the potential value of the solution not being achieved and money being wasted. For example, a company wants to limit storage of emails as they have seen that if 80% of employees do not have personal folders (PST files), the company would save $400,000 dollars a year. They therefore set this process into motion and designed a removal tool to remove them, and then disable this function. They didn't do any testing, assuming that the removal of this would make the employees behave better at clearing old files and only storing what they needed. What actually happened was that employees really relied on the PSTs to back up emails and attachments, so instead they started storing these on external drives. The company encrypted their employees PC's, this unintended consequence meant that there was now company data on unencrypted devices.

Another company created an enterprise social network, thinking that it would be a great product that they could sell on the premise of engaging employees, and people would use it to communicate and collaborate. They went ahead and designed and launched their enterprise social network product. Many customers reported disappointing adoption. They found there was an initial spike in usage as people checked the new program out, however, this rapidly dropped as employees had no time to really use it. Also, they seemed to IM each other as opposed to writing on the wall of their fellow employees. The value the company had seen in the social network was therefore not achieved. Had they conducted some concept testing they would have learnt that while employees may use social networks outside of work, they had little need for it inside work. The only time they saw value was when they could see who new team mates were or what people looked like. The employees also preferred to communicate more privately to enable more open conversation which led to richer relationships, than openly on profiles where they felt they would be more careful in what they said. Had the company done this, they may have changed their product to be a plug in to their customers employee directory with more information including pictures and profiles, as opposed to building a whole system whose value didn't get reached.

With users, as a rule, they **generally do not seem to like change unless the perceived value is high enough for them to switch or adopt a new behavior**. Especially in the enterprise user world, since they are a 'captive' audience as opposed to a 'consumer' i.e. they haven't always chosen the solution. It is a common mistake in thinking 'if we do x, employees will use it, like it etc.' Actually, it has been my experience that employees are notorious for 'workarounds', some even thrive on finding a way around the new system, process or rule that a company has put into place that they disagree with. If you spend all the money, time and resources to create the solution, and then they do not use it... what was the point? Doing these types of tests, can allow you to make informed decisions and get your best possible chances of adoption for the solution. Remember, it is the discussion you will have with the users after these tests that holds the value, e.g. why they felt a certain way.

CONCEPT TESTING METHODS

There are several ways to do concept testing. We will outline just a few here to get you started with thinking about it. As well as the following tests, you can simply use a storyboard and walk the user through the solution. Then, you can ask them about their likes, dislikes, concerns, the value they see in it and any barriers to entry (any reasons they wouldn't use it). You may also need a combination of these tests at different points.

Feeling on a Scale – This is good to see how the concept makes the participant feel, as close to as satisfied as can be or as unsatisfied as can be. It is very good for limited restrictions and general measure. It is worth mentioning, the scale can have anything for the extremes e.g. as productive as they can be, as easy to use as can be etc.

Comparison to Optimal – This is good when you have a known optimal or better solution that you are hoping to match or even beat, building on the feeling scale. You can then see where participants will place your solution concept on the scale vs. the other solution (figure 19).

Spend Test – This looks at what the user would pay or give up to get the solution, or certain features. This is great when you want to look at which features are more important to the participant – especially if you have requirement or prioritization decisions to make.

Bundle Testing – this is great to look at different choice points if you want to weigh up time, features and importance to your target audience; building on the Spend Test (figure 20).

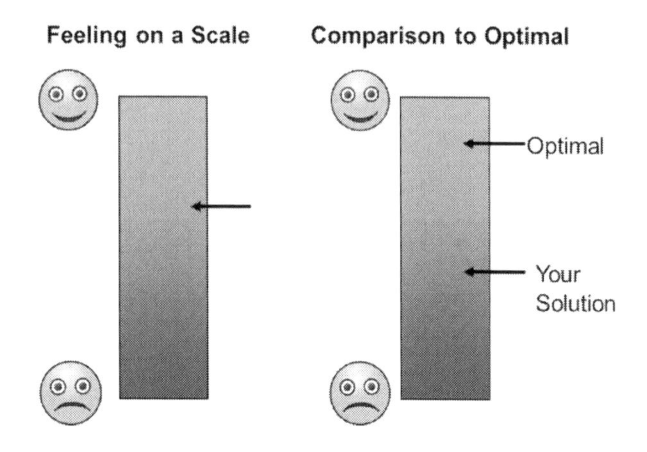

Figure 20 : Feeling tests

PERCEPTION TESTING

Perception testing can be associated with the brand of a solution, organization or entity and is the measure of what people think and how they feel when they have an interaction with the entity. An interaction can be talking about or hearing about the entity, a thought of the entity or actually coming into contact with the entity. Perception is huge in experience, as experience is a space and encompasses everything the user sees, hears, feels and touches etc. Perception can affect how they react to the solution, how easily they adopt it, and what they tell others. Therefore, any preconceived notions are also important to find out to help build a solution they will easily adopt. After capturing this data, you can then feed it into your delivery plans.

It is also important to remember that perception can be changed, fixed or improved. It is needed even more in organizations that are early in their experience efforts, or

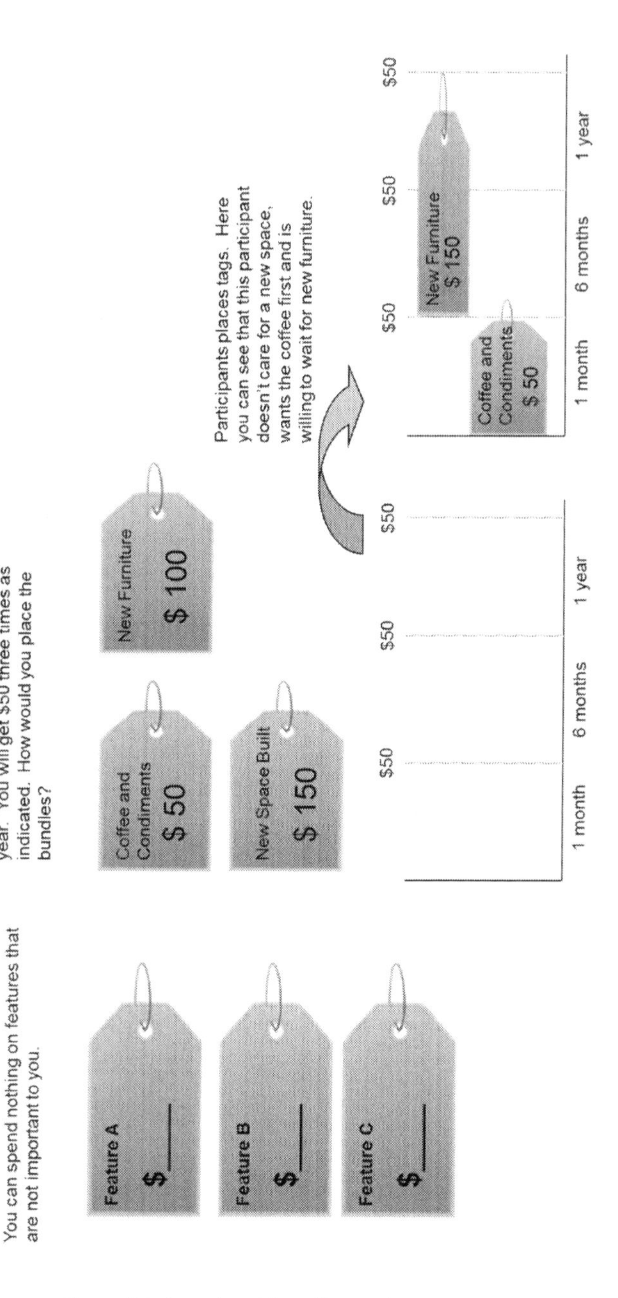

Figure 21: Spend and bundle testing

those that haven't focused on experience design or are not making sales etc. To do this, you have to know what they think at the beginning - what is the baseline? Also, if a user's perception is bad, you have a great chance to win them over and provide an even more positive experience. If done correctly, these converts can become a strong source of positive publicity.

For Example:

Chuck's Burgers had released a new burger last year, they had made a big splash about how tasty it was and assumed people enjoyed it. In actual fact, the burger tasted awful in comparison to Minnie's Burger. Chuck's realized this when sales were down for the burger, so they created an improved version and wanted to release it. They made a splash about how great it was. However, people remembered the original burger, and the memory of this triggered an emotion in them, so they didn't want to try Chuck's 'new' burger. Had Chuck took into account the perception, he could have changed his marketing, he could have even taken the burger off the menu and launched a whole new one.

MEASURING PERCEPTION

Word Association: One quick and easy way to measure perception is to get people's response to the question 'what one word do you think of when you think of x'. You can then categorize this by tagging the words as having negative or positive connotations, and then look further into the themes.

Perception can also be a composite. For example, if you know the variables that affect the experience, then for each variable you can see how they feel on a scale of 1-10 of negative-positive.

For example, if the variables that affect the experience of a burger are (1) quality of meat (2) crispness of bun (3) freshness of condiments (4) choice of toppings, then Chuck could have asked tasters to do a blind taste test on the original and the new burger and rate how they felt their experience was to see if he had improved on his latest 'new' burger.

	Original Burger	**New Burger**
Quality of Meat	4/10	7/10
Crispness of Bun	2/10	7/10
Freshness of Condiments	6/10	8/10
Choice of Toppings	6/10	9/10
Total Experience Score from 40	**18**	**31**

Table 11: Perception testing

It is important to remember in these tests to probe the participant. If you don't know the reasons why they feel a certain way, how will you know how to fix it? It is incredibly important to know where you are starting from in people's minds, so that you can create marketing and communications that will enable adoption.

USABILITY TESTING AND DESIGN

Usability testing is concerned with testing the actual solution itself. It provides insight on a behavioral level. As Jakob Neilson said, when creating for users, you want to aim to 'make it easy to do what they want to do'; therefore you need to understand the task they are trying to complete and how they behave. It is very much recommended that if your role is in usability that you read some of his work – it is a fantastic foundation to the usability world. Usability testing is based on cognitive science and exists for products to be more usable, which has been interpreted as meaning the product can be used with minimal or no explanation.

Given the newer emotional aspects that have come into play with design, to create an experience more is needed than task orientated notions alone. This is because the feelings they have will affect the users want to use it. Their feelings are more important than many people in a less mature organization realize. When in a state of 'negative affect', such as feeling frustrated, the mind narrows on detail. Therefore, things that normally people may let slide, they will now notice and be more annoyed by. In a state of 'positive affect', such as when feeling happy or relaxed, the mind is capable of seeing the bigger picture, so minor details fly past unnoticed. This will affect the user's overall perception.

You want to design the solution to stimulate various emotions as needed, to drive the particular outcome you need. For example, you may want to create an enticing or engaging experience when you want to promote awareness or when you are getting people to try out something. In another case, you may want to promote more of a serious feeling for trust or a feeling of simplicity, for example if you want the user to complete a financial transaction.

The key is that the solution will be effective when it is tailored to the intended task.

As Donald Norman says, 'a good design is when beauty and usability are in balance'. This speaks to both the emotional and functional parts of design. As I like to say, design for '**beauty and function**'.

There are various parts of design that need to be taken into consideration - all should be thought through to make the design and experience of the solution cohesive and successful in its purpose. The art of design is balancing these to fit the goal.

Sometimes people are so focused on appearances that I've seen even high-level executives like something just because it looks 'cool'. However, all aspects are important. The reflective design is important as if the solution frustrates them, then this is what they remember and will associate with the solution. It will frustrate them further and more so as they are being forced to use something they don't want to – for example captive audiences like employees, or users signed up to a company's service for a yearly contract and then the design changes. In this case, if possible they will find an alternative, and at a minimum carry that negative feeling forward about the solution.

Visceral Design	**Appearances** Unconscious and pre-thought. The first impression the user has is instinctive, it is not a preconceived thought and is normally without reason. E.g. reaction to appearance, touch, smell, feel etc.
Behavioral Design	**Effectiveness of Use** Use of the product, its function, performance and usability. Based on how people will use it and what they need it to do.
Reflective Design	**Rationalization of Product** The meaning of the product. This is conscious thought, the highest level of feeling. This is where the overall impression of the product is formed. It is what enables the user to think back and remember the experience. This is where the feeling the solution instilled in them, will allow minor difficulties to be overlooked, or blown out of proportion. This determines if the user would want to return to the product or not.

Table 12: Design types

While we do not focus on the actual visual design and how to do this here, one key point worth mentioning is that just like they say faces that are symmetrical are seen as more beautiful, laying things well on a grid, also makes screens, UI's and other visual designs much nicer to the eye. The amount of times I see UI's where the spacing is off, icons are floating or elements are misaligned I can't tell you. These messy presentations will work against your goal.

You can see how much nicer the layout aligned to the grid feels to the eye. A simple example, but it gets the point across.

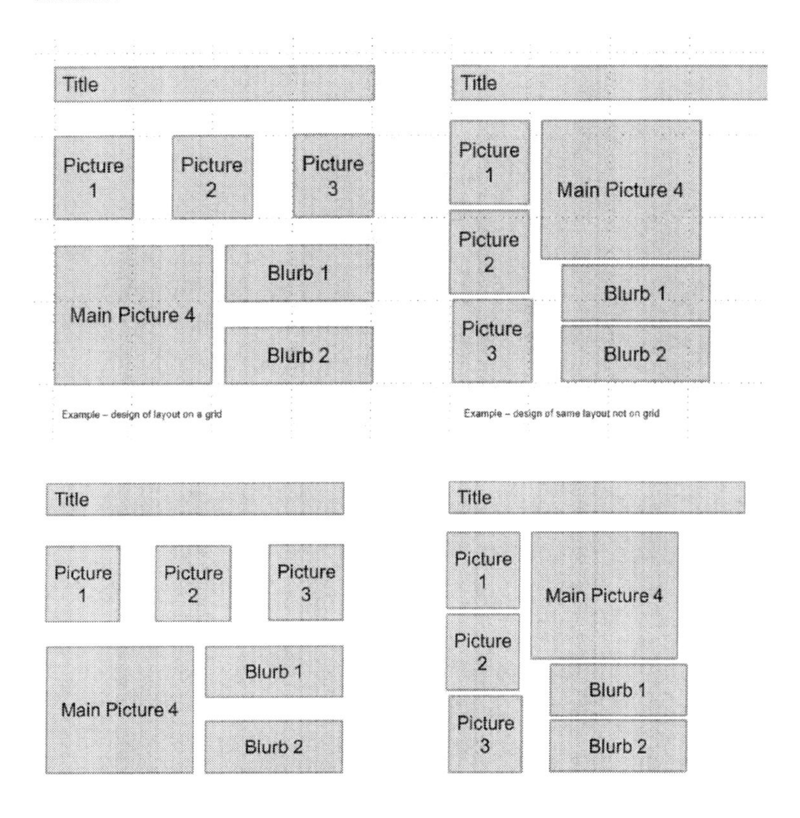

Figure 22: Designing on a grid

Now on to the more tactical information, how do you actually do some usability testing? The easiest way to think of this is: your goal is to test if the users can do the desired task without frustrations and difficulties. You can use walkthroughs of mockups, sketches or prototypes. There are also varying levels of granularity. You could test a large action such as 'make a call', which would include 'finding the number' and 'dialing the number' in sequence, or just the single task of 'finding the number'.

You can use the user flows you created to find the test cases (Chapter 5). It is also good to think of iterative testing, you could have several iterations as you make changes and retest. Usability testing can be very agile and quick. For example you could use sketches of the screens of an online service, and do several iterations as you make fast on-the-fly changes. You could also have two ideas to test against (A/B Testing), and after testing find that in fact the ideal solution is a hybrid of both. To validate you may need another iteration of usability on the hybrid design. Very important for growth and institutionalization of experience within your organization is to prove the value of the discipline. Proving the value of usability testing is one of the easier areas of experience to prove to an organization. You can simply get before and after satisfaction scores or ease of use scores at the beginning of the test and then after. In one of my jobs, I ran a proof of concept for why we should do usability testing. I found across a variety of tests, usability testing improved satisfaction by at least 1.5 points on a 5 point scale. You can also look at the errors you find, when users cannot complete their tasks and ensure that you increase the success rates.

When should you do usability testing? The usual answer here is as early as possible but also throughout design and even in implementation - again, think iterations.

TYPES OF THINGS TO TEST FOR

Some of the fundamentals you may want to test are:

- **How easy is the solution to use the 1st time?** This helps to know how much help people may need to use it the first time. You want to think about how fast people can learn to use it, so think 1st time user as well as seasoned user. This is linked to how often they will use the solution, for example if they use it just once a year they may not learn it well enough to remember it next time.
- **How do users actually behave?** Do they use the system the way it is intended to get the task done? For example, in a website you may have placed a button or link somewhere, and then find they actually look somewhere else for it naturally.
- **Where do users get stuck?** At what point are they confused as to their next action or with what to do next? This can help you identify where you need to give them help or design the solution so that they know.
- **What errors do users encounter?** Is it due to design, lack of information etc.? What is the severity of these errors – can they not complete their task, or does it cause minor or major frustrations?

- **Where do users get confused**? Looking at the choices they have, and thinking of the workflows - do they manage to follow the flow and reach a successful state? How many times do they not reach a successful state and what is the severity of this.

You really want to concentrate on looking at what the user does, so that your design can match the way they think and you ensure they have the right information at the right time. You need to strike a balance in the number, length and difficulty of the steps the user has to go through; this is something you can watch for in your testing.

Understanding the goal of your test is incredibly important. Many times I have heard newbies in the space say 'I am testing the usability of this'... but what does that mean? This is important as it will affect how you set up the test, and the validity of the test. For example if IT is testing that an employee can find a page on the intranet when they encounter an issue, and in the test they place the user on the intranet and tell them to find the page, this is not testing that the user would go to the intranet in the first place. If they wanted to test that the user would be able to find it at the time they needed it, they may have wanted to start the user just with the computer switched on and see what they do as perhaps they just go the internet and search. Placing them on the intranet would be testing the actual site, and if they locate it once they were on the intranet already.

It is worthwhile to note, if you are working on improving an existing solution you may want to conduct a usability test on the existing solution, to understand where the problems are with it.

When you report out your usability tests in a report, you would want to at least include:

- The participants and their demographics.
- The methodology used.
- The test scenarios and measures used.
- The results and findings, including errors and severity of errors. For example, could the user not complete their task, complete it with frustration, or complete it just fine. You can even graph this if you award points to error severities.
- Recommendations to fix the errors – even perhaps some new design ideas.

USABILITY TEST PARTICIPANTS

How Many

This may shock you, but 12-15 people are enough (Tom Landauer and Jakob Neilson). This has always been a shock point to much of management that I have had to 'sell' usability to, who are accustomed to big numbers and wanting X% of the total population to be tested. I have used the following graph (along with the math) to explain to them, and have more often than not been successful in demonstrating the difference in this type of data as well as how achievable experience based design is.

Here is the math Landauer and Neilson discovered to prove this, and I cannot thank them enough! In fact, I use around 10 in many tests when creating solutions for hundreds of thousands, and it works great.

Over a vast amount of studies they showed that the number of usability problems in a usability test with n users is:

$$N(1-(1-L)^{n})$$

N = Total number of usability problems, L = Proportion of problems found with 1 user

Typically the L value is 31%! With 0 users you get 0 insights but even with a single user can find a large number of bugs, then with the 2nd and 3rd, you may see more and repeated bugs, finally with the final participants perhaps seeing some corner cases. The graph below illustrates this.

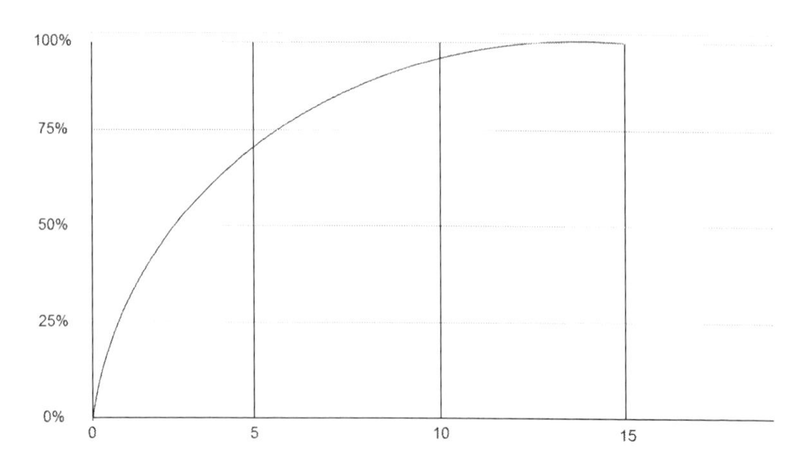

Figure 23 : Graphical representation of number of participants against % usability problems found.

This number may even be shocking to you; however the truth is, the first few people will encounter most of the usability bugs, and then the rest may lead you to some corner cases. Humans, it seems, do not behave that differently and as such, a small group find a lot of the common usability issues. Also, this form of testing needs a small sample size as it is focused on what people do, how they use the solution, and gains deep and focused insights.

The key to remember is **iteration and validity**. You may want to do 3 studies each with 5 participants as you improve the solution design. You also want to make sure you truly are measuring what people do and not what they say. To illustrate this point there is a great example from Wal-Mart who made a $1.85 billion dollar mistake in customer experience. They had surveyed their target audience on decluttering their stores. Their customers responded that they would prefer less clutter in the store. Using this information they rolled out a major change strategy for their store customer experience. As a result there was a large drop in sales. What Wal-Mart had done, was interpret the agreeance to a preset thought or strategy, instead of looking deeply into the customer need. The Wal-Mart customer needed a vast selection of cheap items, not a decluttered store. This is a great example that illustrates that market research can answer questions on general trends and opinions etc. while **user experience research looks deeply into the entire user environment, who the users are, their needs, culture and how everything interacts.**

Who?

There are a few points to consider when selecting your participants:

- **Who is your target audience?** For example, is it all users or a section or particular segment? You will want to be sure you are testing with representative populations. For example, if you are creating an HR tool for administrative assistants to do multiple calendars, and then you test it with sales users, it will not be as valid as if you test with administrative assistants...Duh.
- **What variables can affect their behavior?** For example, country they work in, gender etc... You will want to get some representation from each of these variables in your test group.
- **What variables should be considered in the solution itself?** For example, with the HR system it will be online - perhaps different administrative assistants use different web browsers, that you may want to test with.

With this, you can also try to estimate the amounts in your test group and match the percentages. At a minimal though, you would want coverage of the different variables (figure 23).

Percentage Matching

You are creating a tool for your sales people. You have both male and female sales representatives, they use a variety of web browsers on their PC's including Chrome and IE and they are in 3 main areas culturally – North America, South America, Germany.

You plan to use 10 users in total for usability testing and need to calculate the amount of each you should aim for.

100% of participants = 10

Variable break down for company X:

Variable		Percent in Company	Number Participants
Country	North America South America Germany	30% 30% 40%	3 3 4
Gender	Male Female	20% 80%	2 8
Web Browser	Chrome IE	20% 80%	2 8

Total of 10 Participants.

Matched as close as possible to the variable percentages for the target audience breakdown.

	Male	**Female**
North America	1 - Chrome	2 - IE
South America		1 – Chrome 2 - IE
Germany	1 - IE	3 - IE

Figure 24 : Participant percentage matching

WHAT ARE YOU MEASURING?

You will want to know their baseline and thoughts or perception of the existing solution if there is one, or of the topic in general if it is new.

During the test you will want to measure against your objectives. For example, if you want to test if the user can locate a page within 30 seconds, then you will want to ensure you time that scenario. You will want to note both quantifiable information e.g. number of clicks, right or wrong etc., as well as qualitative information such as when the user displays signs of confusion or frustration.

TESTING METHODOLOGIES

You will have some decisions to make when designing your test - all with pros and cons.

1. **The Level of Assistance** - This is whether you are actually present or not, as well as how much you will interact with them during the test. Pros include that you will see it first hand and can also probe further when you need clarification or deeper insight as to why the user acts in a certain way. The main negatives here are that your presence may alter their behavior; they may act how they think you, or society would want them to, as opposed to how they actually would (social demand).
2. **The Level of Guidance** - This is how much you dictate what they should test. For example if you just let them 'play-around' on the solution freely,

or you list the tasks they should attempt. The pros here are the fact you can choose the tasks to ensure critical tasks are tested; the con is that they may be doing the task, but not in their natural environment or situation.

You will want to create some scenarios for task based tests that represent the situations a user would be in when doing the action. For example, if you want them to test the search on a social network you could say 'you need to find an expert in marketing, how would you do this?'

You will also need to pick the way you want to gather the information - whether it is by survey, focus groups, interviews or observations. Each has their pros and cons, mostly surrounding time, labor intensiveness and richness of data.

Surveys are the quickest method in comparison, as you spend time developing the survey but the user can complete it without your presence, and you can analyze the results later. That being said, survey data is not as rich as other forms of capture. You don't see what they do, and sometimes people just fill it out quickly, remember the Wal-Mart mistake!

Interviews are more time-intensive than surveys, however they provide richer data as you can probe further as needed. If you conduct them in-person, you can analyze a person's behavior and reactions while they answer you - which can provide valuable insight into how they are feeling.

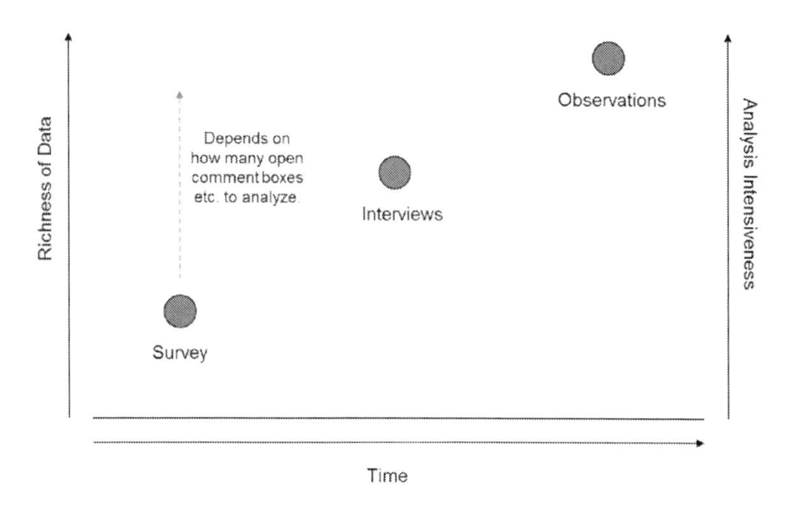

Figure 25 : Different methods comparison

Observations are the most time-intensive, however they give incredibly rich data as you can see what users actually do, not just what they may say or tell you. Observations also allow you find gaps in the experience, or user pain points that you can then fix.

In focus groups usually the solution is demonstrated to the users (as opposed to the users actually using it themselves) and then discussion is held.

More often than not, especially in leaner or newer experience organizations, I have found combining methods works well. For example always ensuring at least 3-5 observations, then a survey, focus group or interviews to validate. Even when writing a survey, it is good practice to get a couple of people to respond to the survey first, to ensure the questions are interpreted as intended.

Regarding logistics, especially for a global user base, you may not be able to only do in-person testing and may require virtual tests. Observations can still be done via desktop sharing software. In this case you can ask participants to share their desktop screen, use a webcam or talk out loud at a minimum.

TEST STRUCTURE

There are four main parts to the test.

1. **Brief the Participant** - At the beginning you will want to tell them what the test will involve e.g. I will ask you to perform some tasks, and then we will have a short discussion. Here, you want to be careful not to give away any information which may make them more aware in a test. For example, if you are testing if they can find a new feature on a webpage and the feature is called 'Alerts' - you don't want to say to them before the test 'we are going to see if you can find the Alerts', as they will naturally now be aware of what they are looking for.

2. **Collect Demographic and Pre-Data** - Then you may want to gather any information related to variables that may affect their use of the solution (or send a pre-form prior) e.g. language they speak, location, gender, what browser they use...this, of course, is dependent on the test.

3. **The Test Itself** - this is the scenarios you want them to carry out, or related questions about the solution, depending on what test you are doing.
4. **Debrief** - this is where you can have open discussion with them, let them know what you were actually testing for if you couldn't in the brief and ask them their general feelings about the solution.

METHODOLOGY 101

A key point to remember on the methodologies: Much of this may seem like common sense, but I have seen many people new to this space carry out completely invalid tests as in the moment, they forget these fundamental points.

Observations

These test structures take time but give amazing data as long as they are carried out in a valid manner.

You should write everything the user does, and analyze it all after. Some people even choose to video the observation so that they can look back later without time restrictions. I myself have watched people analyze the participant's actions in their head and write down their assumption of why a participant behaved like they did, as opposed to what actually happened. I have also seen observers say 'it's right there' when a participant couldn't locate something, then the observer wrote down that the participant was able to locate it! You need to be objective, so the easiest way is to write everything they do then look at the data after.

A way I have found helpful to teach people how to be observers is to allow them to run the observation while I also observe quietly. I write the notes I would take had I been the observer running the test, then afterwards I compare my notes with theirs and show them where they may have been subjective. Sometimes, if the whole organization is new to the space, just having two people observe and compare can help achieve more objective data. There is a learning curve to this skill, I have seen it take anywhere from ten to many observations for someone to leave their own thoughts out of the data capture.

Interviews

An interview can have as much or as little structure as you want, which is fantastic as it allows you to probe further when needed to really find out the core reasons and thoughts of the participant. You want to remember what it is like to be that inquisitive 3 year old, that irritates every adult with the question 'why?' For example, if the user says 'I didn't like the new system', you would want to ask 'what didn't you like about it', and then you may want to probe further with questions like 'what would have made it better?' Or if you want to see importance 'which of these were the most annoying to you?'

You need to be careful not to lead the participant into an answer. I have seen interviewers new to this space often phrase the question so that the participant is more than likely to answer in a certain way. For example 'what issues are you facing with product x?' Who said they had any issues with it? Perhaps now they will think there are issues. Or another example I see a lot of, is finishing with an agreeable statement such as 'the new feature is great isn't it?' Naturally the participant is more inclined to agree - when maybe that is not how they feel at all.

Again, in interviews, objectively scribe what the participant says. Many interviewers use a recorder sometimes to ensure they do not miss data and so they can go back and clarify points. Today there are all sorts of technologies available, such as voice to text.

I have also experienced people get rather upset and defensive with the usability results. One thing I stress to them is the need to just put their feelings aside and have a thicker skin, as they should be focused on what works for the user and not their personal preference. This happens especially when the designer of the solution is the one conducting the usability testing. Understandably, the solution's design is the creators 'baby' so they are more attached to it, however, sometimes in organizations just starting out to focus on experience, the designer is also the one doing the testing. In fact, it is good that they actually see first-hand users with their products, so even if an expert is conducting the tests, it is good for them to sit in one or two (silently).

A good key to go by is think to yourself - how can I act on this data? This is a good indication of when to probe further - for example if the user just says 'no', then how can you act on that without finding out the reasons why?

If you are testing in-person, it is also good to note down signs you get from what the participant is not saying and from their body language. For example, if you ask a question and the participant starts sounding frustrated, perhaps crosses his arms, then maybe the experience was more negative than he is letting on.

Surveys

Surveys seem so simple, and it is my experience that many organizations feel it is about throwing together some questions, yet many surveys I get are what I would call invalid.

It is good to select the right question type for your purpose, and is generally good practice to have comment boxes for main sections to enable the participant to tell you more than the question allows. One comment box at the end is better than none at all, however it will take more analysis as the user may write comments that addresses many of the questions.

Here are some of the main question types:

- **Choice** - This where you give them a set of responses to choose from, you may want them to be able to pick multiple or just one. You can have an 'other' box if you want to see what their true choice would be (in case you do not have it listed). However, you may want to 'force' them to choose and have no open box, as then you will get to which one they would choose if those are the only options feasible.

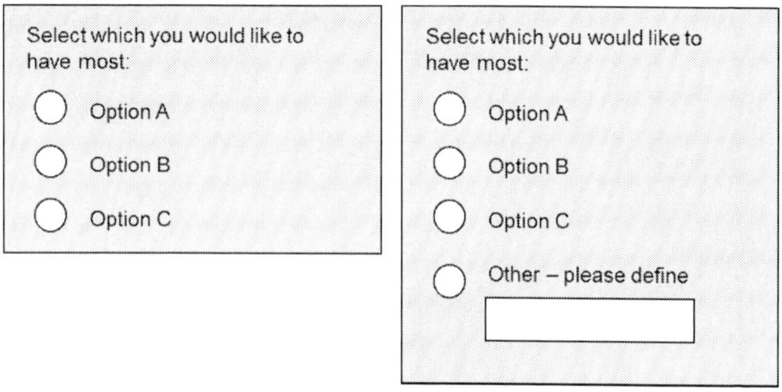

Select which you would like to have most:

- Option A
- Option B
- Option C

Select which you would like to have most:

- Option A
- Option B
- Option C
- Other – please define

Figure 26 : Forcing a choice or leaving it open

- **Rank** - This is where you get the participant to order something e.g. list the following three items in order of importance to you. This gives you some sense of order. However, here you must consider what happens if any of the items are of equal importance. You can leave an open box to ask why they ranked the items this way to give you further insight.

- **Scale** - This is where you ask the participant to select how they would rate something on a scale. Popular scales are 3, 5 or 10. 10 is quite a lot, unless there is a clear difference between say a 3 and a 4. 3 can sometimes be restrictive, both 3 and 5 allow for 'middle ground' where as you may want to force them to be on one side or the other. You need to think about the purpose and goal of the question. Another point to consider in scaled questions is the use of words or sentences as opposed to numbers for example:

How easy was the product to use on a scale of 1-5 where 1 is very easy and 5 very difficult?

Or:

Which of the following statements do you agree with the most?

- I could complete the task with no issues.
- I could complete the task with some issues, but I overcame them quickly.
- I experienced some issues with completing the task and was not able to overcome them quickly.
- I could not complete the task

The thing with numbers is, they are open to interpretation - a 4 could mean one thing to one person and something else to someone else, therefore in some cases using a word or a sentence to state what you mean hides the scale, but is still a scaled question to give you that level that you need.

Once more, you should check your survey by asking yourself, how can I act on this information if I get a response of ‘x’. For example, if 40% of your participants say they found the ease of use 2 on your 5 point scale - what can you do with that? If the survey is just a pulse or for giving high level sentiment then that will be ok, as you can then follow up with interviews etc. to get a deeper insight. However, if the survey is your only data capture, you may want to consider open boxes for the participant to explain their choice.

Something else to consider, is that getting people to respond to surveys can also be hard. You will want to ensure you give them a rough estimate of how long the survey will take them, a deadline, and even consider incentives.

Comment or open boxes are worth another mention as they can provide a deeper insight for the user to explain themselves further. If you only have a small number of responses, you can afford to have more open boxes as you can do the analysis easily. With many responses data analysis of open text is laborious (although there are an ever increasing amount of very good text analysis tools on the market).

Survey Checks

There are a few rules of thumb to consider for your questions also:

- They should not be double-barreled. For example 'was the feature easy to use and understand.' If they say yes, which was it, or both? Perhaps it was easy to understand but they actually did not find it easy to use.
- Do you know their frame of reference? For example if you are testing an improvement to a system - are they the user of the existing system or not? As an existing user will be comparing it to the current system.
- Spaces for validation and insight - have you put open boxes to gather deeper insight?

- The questions should not lead the participant. For example, 'the search was really easy wasn't it?'

Survey Data Analysis

For some types of information the way in which survey data is analyzed is extremely important. It is not the same as getting a great quantity of responses and just summarizing a column to say 80% said yes. Instead you want to think of it as reading across the row to get a real feel for the person, their actual thoughts and mental model. If you read a survey and cannot be sure what the participant meant, if possible get clarification, this is better than attempting to guess what they meant - as you may be wrong in the conclusion you draw.

For example let's take the HR career site. We want to know if people would use it and how we should market it.

We want to include a variety of question types to get a general sense and also look deeper into the participant's thoughts. Below you can see the questions and some responses we got from 3 participants.

From the first question it looks like everyone would use it... but is this really the case?

Would you use the Career Site? (yes/no)	Are you currently looking for a new role? (frame of reference/ context)	Do you tend to use Monster or other career sites when you are looking for new opportunities? (mental model/ tendency)
Yes	No	No
Yes	Yes	Yes
Yes	Yes	No

Table 13: Survey data

- Participant 1 - isn't currently looking for a new role, so may not have interest in it right now. They do not tend to use career sites in their 'outside of work' life anyway, so they may need encouragement to use the tool inside.
- Participant 2 - is likely to use it, not only do they say yes, they have a tendency to use similar sites externally - they could be a potential ambassador for the site when you deploy it.
- Participant 3 - while they are looking for a role, they do not tend to use these types of sites - you may need to convert them with the benefits given their current need.

You can see how just reading the 'yes' does not give that deeper level of insight, but reading across the row of responses, tells you a little more about the person. Of course, this is a very simple example, without open boxes and at a high level - but you get the gist.

Now, when you add demographics it can help analysis and insight further. For example if we look at age ranges, perhaps participant 1 is 50-55, 2 is 20-25 and 3 is 50-55. This type of data may help you further understand the minds behind the responses. You then can really craft your delivery plans and launch to suit the users and get as much adoption as possible from the start.

Survey Comment Analysis

As mentioned there are many tools for survey comment analysis coming to market that can do all sorts of analytics on text. Given human nature, I still find that a person actually reading the comments, reading all the responses of an individual and understanding who that person is and what they are trying to say is incredibly valuable. There is a method we use often that we call 'tagging' to allow us to get the ever important number and measures from comment fields. It is very simple, but the key is in understanding your tags, and how the comments are interpreted.

Let's take the following comments from a survey:

How was your installation experience?
Awful – the process was slow and at one point I didn't know if it was broken or not.

I called support pretty fast and David helped me very quickly and professionally.
Really simple, didn't really do much just hit next on each screen.
OK
Not good at all, I expect more from such a high quality company.
Horrible
Slow
OK – it could have been less steps and faster

Table 14: Survey comments

Step one – start tagging the responses, first tag it with the overall sentiment positive, negative, neutral or mix. Then tag it with the main category and sub category. I find having at most 2 categories is enough – as these are the main thoughts that the user was having.

Table 14 shows what this would look like for the small data set above for the first 4 comments.

How was your installation experience?	Positive/ Negative	Category 1	Category 1 – Sub-Category	Category 2	Category 2 – Sub-Category
Awful – the process was slow and at one point I didn't know if it was broken or not.	Negative	Process	Slow	User Feedback	Progress Bar
I called support pretty fast and David helped me very quickly and professionally.	Both	Error		Support	Saved Experience
Really simple, didn't really do much just hit next on each screen.	Positive	Process	Simple		
OK	Neutral				

Table 15: Survey tags

Once you get a feel for some of the main categories and their subs, add these to a tag set. Then, when you continue tagging the data you can see if these tags apply.

This starts creating the trends and themes in the comments. If a comment doesn't fit any tags in your tag set, add the new tag to the set.

Continuing with the example comments -

Our current tag set would look like:

Main Category	Sub Category
Process	Slow
	Simple
Error	
User Feedback	Progress Bar
Support	Saved Experience

Table 16: Tag categories

Then we continue tagging. As you can see 'process -> slow' comes up again and again. We needed to add a general 'bad experience' tag and a 'too many steps' sub-category for 'process'.

How was your installation experience?	Positive/ Negative	Cat. 1	Cat 1 - Sub	Cat. 2	Cat 2 - Sub
Awful – the process was slow and at one point I didn't know if it was broken or not.	Negative	Process	Slow	User Feedback	Progress Bar
I called support pretty fast and David helped me very quickly and professionally.	Both	Error		Support	Good Experience
Really simple, didn't really do much just hit next on each screen.	Positive	Process	Simple		
OK	Neutral				
Not good at all, I expect more from such a high quality company.	Negative	Unmet expectations			
Horrible	Negative	Bad Experience			
Slow	Negative	Process	Slow		
OK – it could have been less steps and faster	Negative	Process	Too many steps	Process	Slow

Table 17: Tagging

Our tag set would now look like:

Main Category	Sub Category
Process	Slow
	Simple
	Too Many Steps
Error	
User Feedback	Progress Bar
Support	Saved Experience
Bad Experience	
Unmet Expectations	

Table 18: Tag set

Using the tag set also helps if you have many survey taggers working on the same survey. Once you have your tags – you can obtain the quantitative data as well as having the qualitative data you need for the core understanding.

You can use this in your reports, for visualization such as the data displayed in figure 26.

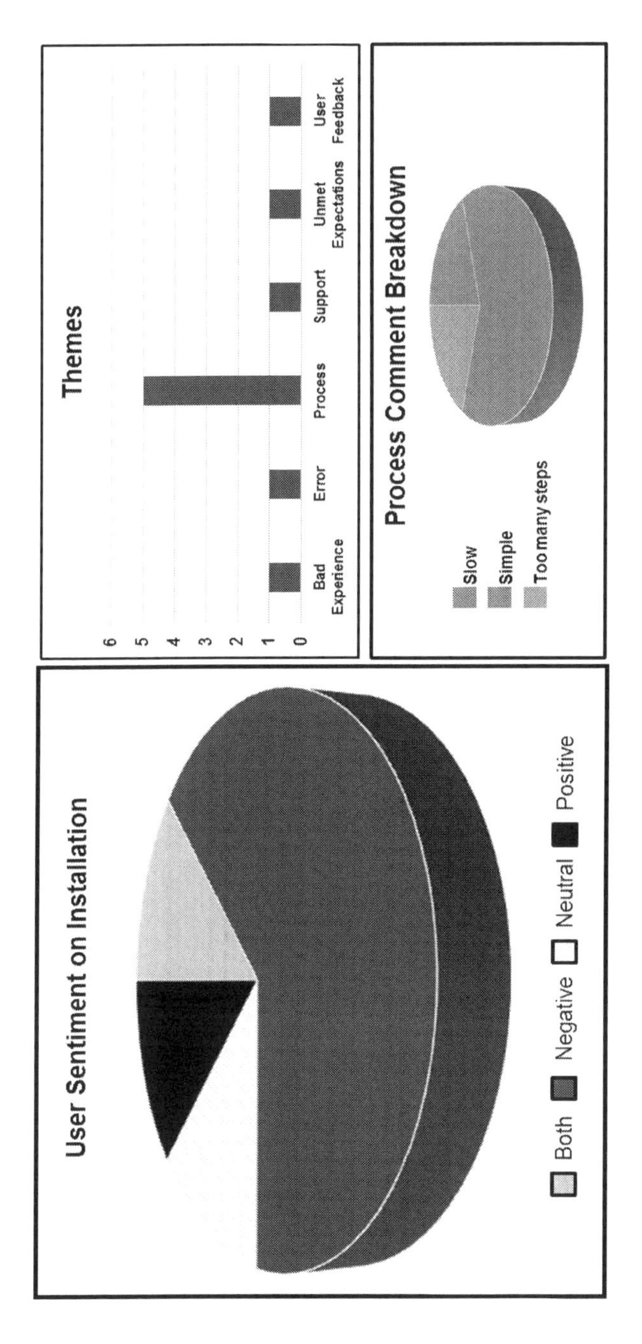

Figure 27 : Example data you can use to report out

FEEDBACK

Feedback is often a very broad term and can literally be any time a person tells their thoughts on anything. While what you learnt in usability testing is one type of feedback regarding the testing of the actual product, concept testing is feedback gathered very early on, feedback is also something you may want to collect on delivery and something you will want to collect on an ongoing basis. You will want to use some mechanisms to collect user feedback that can be incorporated into next development cycles, next generation products or improvement plans. It is this feedback and ongoing analysis that is fed into the strategy and roadmaps to bring the experience thought process full circle (reference the ROAD™ framework).

Here we will discuss some of the main concepts, types and mechanisms. Remember that the data collection methodologies can all be used at this stage for gathering feedback e.g. interview, survey, observation. It just depends on the goal and intended use of the feedback.

The general rules on data still absolutely apply here:

1. Validity: right people, right data, right time.
2. Wrong feedback = wrong decisions.
3. Design how to capture the feedback based on the GOAL of the feedback, whether it be open or focused.

There are at least three types of feedback for you to consider capturing:

1. Check how delivery of the solution went.
2. Ongoing capture for how the solution is doing and trending.
3. Spot check, pulse or audit to check on the experience the user is having at a given time.

Feedback on Delivery

You will want to consider how delivery went – depending on how you chose to deliver the solution. Perhaps a few days after a user downloads a service or signs up for an account you may want to see how he is doing with the product with a survey or a customer call. If there were communications and training as part of delivery, perhaps you will want to ask questions about the adequacy of them, so that you can make improvements where needed before you deliver it again.

Ongoing Feedback

There are several ways to capture this, whether it is interviews every couple of months where you can ask very focused questions that may hold relevance at the time, such as the month a new feature was introduced. Or, perhaps you have a feedback link embedded into the solution where users can leave their comments and feedback whenever they feel the need to. The pure nature of feedback in this way sometimes leads to a skew or bias in the data – as we find that people are more likely to leave feedback when they are angry or dissatisfied. You will want to determine the frequency you capture feedback, how you will capture it and demographics needed in the data set.

Spot Check or Pulse

You may want to do periodical observations to really see how the experience is. Observe your users in the 'field' so to speak. See how they are using the product in their roles and capacities and then analyze this to see the ongoing experience trends. Perhaps you do a spot check before the next major project to get current information from real end users. Another great spot check, especially for service orientated experiences is the 'secret shopper' style of testing. This is where someone uses the service or product 'secretly' testing it. For example, the manager of a Café wanted to know how his customer experience was, so he got his friend to go to the Café and order a coffee. He then interviewed his friend on the keys to his experience (again thinking of those variables defined from the roadmap and strategy phase). Here for example he asked if the friend was greeted on entry, if service was fast enough, if they felt welcome, if they were offered food as well as the quality of the coffee and cleanliness of the environment.

Today, people's thoughts on products and solutions are everywhere, especially given the wide use of online social platforms. It is wise to bring together your online reviews and scores (such as 'likes' or 'recommendations') as a part of your ongoing feedback review process.

7 DELIVERY OF THE EXPERIENCE

The stage of delivery to paying customers, when the decision of what to buy is entirely their choice, is generally taken extremely seriously by companies. Much time is spent on analysis, creating the product and campaigns to launch the product, taking every measure to increase the chance a consumer selects their product and that sales are maximized. Before, you may have seen a commercial on TV and decided to go to a shop to look at a product, then buy it. Today, anything you tell a user, can be searched. Users have the ability to quickly go online and see if other people confirm, recommend or like a product or company.

It is a common misconception that if a company gives users a service they will just use it, or if a company tells them something they will just do it... but two things arise here to point out. One - remember assumptions and how dangerous they can be. Two - thinking of users as an audience that will do as you say, or use what you give, is also dangerous. Just because someone has bought your service, product or application does not mean they will continue, recommend you or be a repeat user.

Employees may not have a decision of where, and on what, to spend their money like consumers. However, everything I have seen indicates they are more than willing to create a workaround to something they do not like or something that is not easy nor intuitive, which in turn can cause companies unintended consequences. This makes the fundamentals of delivery from an experience perspective pertinent to everyone, as in this world everyone has choice. Take for instance, Company X realizes that employees are using a well-known application for listening to music on their company provided PC's. They then issue a policy stating that this product is not allowed to be installed with no information to employees as to the reason of why this is important. Company X disables the product from being able to be installed, assuming that employees will therefore not use it. The Networking Team notice that things are running a bit slower than usual, files are loading slowly and people start complaining. Turns out, employees started using an online streaming product to listen to music as they work, causing the unintended consequences the company experienced.

The equivalent measurement to seeing value of comparable relevance to that of sales, is adoption. When employees use what the company gives them, or carries out the intended action, then the company sees value. The reasons a company delivers a service or product, or says that employees should do a certain action, can be for many reasons including compliance, money savings or productivity, and as a by-product satisfaction can be affected. Consumer side usually focuses on sales, however, they should also be looking at adoption as well as experience measures.

To deliver your solution to an audience is no simple task. Many companies subscribe to the soul wrenching pattern of email them once, twice and again in an effort to make them use the solution or carry out the desired action. This is synonymous to the teacher who tells a child something and the child doesn't understand. The teacher then repeats themselves louder, as opposed to changing the way in which they deliver the information to the child - e.g. visually instead. The same is true to user delivery. You must, **each time, plan the best way to deliver the solution to the intended audience.**

It is absolutely true, that every time you deliver a solution to a user, they may not want it nor like it. In companies there are many times when this may be the case. It then becomes a question of how do you deliver it so that the user experiences minimal impact and can continue working. Ensuring this, will give you maximized adoption allowing the company to receive the intended value.

The key to success here, is to understand:

- Some fundamentals of human behavior.
- The change that will need to occur due to the delivery, from their perspective.
- How to maximize adoption.

First, we shall take a moment to look into some keys about human behavior. Knowledge of this will only help you in your endeavors, to increase your credibility and help you navigate and achieve change in human behavior.

HUMAN BEHAVIOR

Human Behavior (B) has been described as a purposive reaction of a Human Being (P) to an idiosyncratic meaningful situation (S). This can be described as Behavior being a function of the person, and the situation: **B=f (P, S)** (Dr. Uwe Dompke, 2001).

In other words, the differences you see in how people behave, is attributed to the characteristics of the person themselves, and the differences in the situation they are in. Simply, human behavior is a goal oriented reaction to an observable stimulus (something prompts it) with three components that will determine it - cognitive, psycho-motor and socio-affective.

Behaviors are also dependent on interactions with other people, whether the person is in a team, group, crowd, or what the public opinion is. Especially in today's world where everything is at your fingertips. Information and social media play a huge role in defining how people will behave, what they will expect and how they will react.

There is a definite collective nature to human behavior. Since the beginnings of time, people have grouped together in some manner, and now, with social media, we can be collective physically or virtually.

It is extremely easy for your mind to race as you think about human behavior, how many people you have to deal with and how many ways they can behave. Suddenly, you are imagining a delivery plan tailored to each individual. However, to be honest, humans are just not that different in the way they will behave. This was a key learning point for me, and a key piece of information that helped me educate others.

Across a broad range of situations, network effects (people networks e.g. the groups they are part of) can predict 40% or more of the variation in human behavior. Think about the power of this. The social element of today's virtual world and environment is far reaching, and you must consider this. The point is that humans display predictable behavior, according to those in their networks. Brown (1991) actually compiled a list of about 400 traits shared by all known human cultures.

So remember, **human behavior is much more predictable than is generally thought, and their behavior is affected by the behavior of those in their networks.**

Picard (1997), Ariety and Lowenstien (2005), also found something important to remember - when angry or fearful, humans rationalize their behavior. For example, say the users are fearful of installing a program that they do not understand why they need to install it and then choose not to install it, they may think that they are justified in doing so. It was also found, that humans displayed 'patterns' of behavior for example a weekend pattern, a socializing pattern, a waiting pattern etc. This further demonstrates the limited variability in behavior.

So remember this key nugget, that **the behavior of most people is likely to be regular and predictable.**

Now, the task of understanding people and delivering to them in a way which drives them to behave in the desired manner, doesn't sound so daunting does it?

DESIGNING TO PERSUADE HUMAN BEHAVIOR

There are some fundamentals that the Fogg Model for Persuasive Design talks about, that are key when thinking through how to drive humans to display a desired behavior. You should spend some time reading about this model – it definitely helps a lot in many MOC considerations.

The main point, simply, is that you should *T*rigger the behavior when the person is *M*otivated to behave that way, and has the *A*bility to do so.

Trigger + Motivation + Ability = Desired Behavior

However, motivation and ability can have some trade off in the equation. Basically, people need some non-zero level of each. For example, if motivation to complete some task is low, but you have increased the ability to complete the task by making it a really easy 1-click, then they are likely to do the behavior. The level of motivation and ability can be manipulated, you control the way solutions are designed and delivered - therefore you control how easy it is to use and how much you motivate them to use it.

Triggers and timing are usually key, and have some critical success factors. People have to:

- Notice the trigger.
- Associate the trigger with the behavior.

- Have high motivation and ability at the time of the trigger.

Let's illustrate with a quick example. Think how many times pop-ups get used in technology as a trigger for a desired action. For example, a user is browsing quietly on a website and a pop up comes up to say - do you want to buy something with some chat-bot to a sales agent. If the user is motivated at that point to buy and they have the ability, i.e. in this scenario they have a want to purchase something, and the funds to do so, then the trigger is welcome. However, if they have no or low motivation as they are just browsing, then the pop-up can be distracting and a nuisance. If they want to buy, but do not have the funds right now, then the pop-up can be frustrating - as it 'rubs it in their face' the fact that they want to do something but can't at this time.

There are a few different types of motivators that the Fogg Model describes:

- **Pleasure/Pain:** Result of this should be near immediate, with little thinking or anticipation. E.g. doing the behavior will cause the person immediate pleasure, or stop the pain.
- **Hope/Fear:** The fact that the person thinks something will happen if they do/do not do the desired behavior can drive them to do it. This can be used to motivate people, and/or manage expectations and knowledge. For example, if you do not tell users what they can expect, their expectations can run riot, and then they could

have fear that is unwarranted, or a hope that you can never meet.

- **Social Acceptance/Rejection:** People are motivated to do behaviors that give them social acceptance. This goes back to what we mentioned on human collective behavior and social demand. The fact that people tend to do what they think society would do normally, and what society expects of them. The fact is, most people worry about what others think about them, how they 'look' to others, and innately want to be liked and accepted by others.

Ability focuses on how easy it is for the person to do the desired behavior. It is not about teaching them and training them always, as sometimes they do not want to take trainings, and in fact this can be a barrier to them. This is because training or learning something new requires effort, and humans are naturally lazy.

Ability to act in a certain way, has some factors to consider:

- Does it require little time, or enough time for the perception of the task?
- Does it take not much effort?
- Does it cause low cognitive load i.e. do they not have to think too much?
- Is it close enough to routine? Or way out of their 'norm', with no motivation?

These factors are tough and will vary from person-to-person.

PERSUASION BY INPUT STIMULUS

First a few notions we have learnt about (and a few more to add), grossly simplified I may add:

1. The subconscious mind, does not really know the difference between reality and non-reality.
2. We use sensory input to determine the world around us.
3. Our brains use this input to draw conclusions, thoughts and react.

As I said, grossly simplified. An example of this that I saw on TV once is where a chef cooked a delicious pasta dish. Then, for one participant they colored the pasta black and made the topping thick and gloopy, for another they left it looking nice and for a third they gave them the bad looking pasta but blindfolded them so they could only smell and taste it, but not see it. They asked them to rate the pasta for taste. Long story short, the sensory input of seeing the black pasta was enough for participant to conclude the pasta was bad, when in reality it tasted delicious.

Therefore, if you give someone the correct sensory inputs, the mind will come up with the thought. Working backwards, you can think what you want someone to do, then think what they need to see, hear, and touch etc. to make them draw the conclusion. For example, if you wanted someone to conclude that you are a good designer you could provide inputs such as recommendations from peers (hear), portfolio of good work (see) etc., allowing them to draw this conclusion themselves. Isn't that powerful!

Now we know a bit about human behavior, let's look at some keys in change management. When you deliver something, it normally means a change, either a change to a new behavior or stop a behavior. It is some change to the users 'eco-system'; therefore Management of Change (MOC) becomes very important.

MANAGEMENT OF CHANGE - MOC

Whenever your delivery requires some change, especially where it may not be a choice, you need to proactively transition users and plan for their adoption or minimization of impact. To get effective results, i.e. adoption or sales, and therefore the value you need, the quality of the solution and the user acceptance of the solution both play a part. MOC practices have been studied and defined long ago, and while the fundamentals still hold value, MOC in today's world has also evolved. Given new technologies and a world always on and connected, new expectations have arisen and have to be considered. Unfortunately, where I have seen MOC fail is where the implementation of the discipline is treated like a list of checkboxes and every delivery is treated the same. There is no interpretation of what is being delivered, and the people it is being delivered to.

Some traditional MOC approaches we see include:

- Tell them to do it, and repeat it over and over again.

- Get others to tell them.
- Make it a requirement as a measure of their performance.
- Make is a matter of compliance.
- Have punishment if they don't do.
- Create and socialize the vision.
- Inform them why and expect that to be enough.
- Incentivize the action - offer them freebies.

While some seem ridiculous, some will work in some cases. For example, numerous times I have seen freebies work miracles, and how many times at conventions or conferences do you go to a booth or table for that free pen, that you don't even need. The truth is, MOC needs some careful thought to plan for. The problem is, most of these approaches are reactive initiatives in response to lack of sales or adoption and can have multiple unintended side effects e.g. money spent, increased dissatisfaction etc. Also, MOC doesn't mean a ton of extra time. I point this out, as if you are trying to get some of these practices into your team or organization, it is important that others need to 'buy-in' to it and do not see it as a fluffy topic that adds time to their delivery. Instead the point is that in the same time it takes to do a bad delivery, you could have put some thought into it and implemented a good one.

PLANNING FOR MOC

Start with two key points, 1) what are you trying to create as the behavior? A new behavior, do more or less of a behavior, or stop a behavior, and 2) for what length of time will they need to do this behavior? One time e.g. setup or signup, for a defined period of time, or always from now on.

We will look at the Fogg Behavior Model to illustrate some examples to get you thinking along these lines. Remember, this model states that to get the behavior to happen, you must trigger it when the person is motivated and able to perform it. In terms of delivery, this makes you think of when you trigger it, how you trigger it, how have you enabled them to be able to do it (part in design, part by what you tell them), and how have you motivated them to want to do it?

A new behavior just once, such as register their details online can also be used at the beginning of introducing a more complex and prolonged behavior. In this example, perhaps getting them to do this once is actually the first step of a plan, for them to use an online tool moving forward, a behavior that you aim to make more habitual.

The issue with behavior that you want to occur one time is that since it happens just once, the person should have enough knowledge to complete the action; otherwise they can get easily frustrated and not bother. Since it is once, they likely won't want or be motivated to take a ton of training for this one time affair. They also won't have the 'ongoing learning' that occurs with actions people do multiple times. Therefore with these one-time actions, you need to ensure they have everything right there, quickly and easily to complete it. As per the model, if motivation and ability are not there at the time of the trigger, the behavior won't happen. Here you may want to motivate them with an incentive, perhaps those that enter their details by a certain date will be entered into a draw, or you may want to highlight the benefit of the action. For example, telling them that if their details are online then changes will be automatically updated and immediate saving them doing the manual process of filling in and sending an HR form and awaiting someone to log the change i.e. some value to them. To increase their ability, you would want to be talking and communicating in terms that they will understand, as if they do not understand it, why would they assume they could do it?

If the behavior is already known and they are accustomed to it, but you want them to do it at a particular time, it is easier because they already have the knowledge on how to perform the behavior, and have some knowledge on the cost/benefit of doing the behavior. However, you need to consider if they have the ability to conduct the behavior at the time you ask them to. Remember the shopping example? Shopping is a known behavior, so we do not need to teach them how to shop, however having no money means a lack of ability. In this scenario you will need to focus on the timing of the trigger to be when they are motivated and when they are enabled to in terms of the prerequisites. A big challenge is the 'when' for the trigger. In the example, popping it up for someone after pay day - may have better chances at increasing ability. Looking at the motivators, sensation (pleasure/pain), anticipation (hope/fear) and belonging (acceptance/rejection), you also want to see how to manipulate these. For example, if some action that they are accustomed to is being replaced, this could cause frustration, however, if the new action is better than the last, quicker, or easier, the motivation could be the reduction of the pain and frustration caused by the existing action. Take the new online system for storing user details - the users know the current method, and are comfortable with it, they know where to get the form, how to fill it in and who is making the changes. However, it does require a lot of effort and 5 days for a change to go through. With the online system, they have some fear of entering details stored online. You can use the fact that changes will then take less time and effort as it will be immediate to show benefit, and demonstrate the safety of details online in your delivery to get them to perform the desired behavior.

You may want people to increase how much of a behavior they do for a moment in time. For example, you may want your insurance agent to sell more insurance for a day. This type of change is often a stretch for users, but this is the whole point. Through pushing them to do more, or intensifying the behavior for a finite period of time, they in turn push themselves to do it and can therefore see what is possible. The benefits they receive from intensifying their behavior can then become the norm. Here, you would want to couple the trigger with something to motivate them, such as an award, and then you make the behavior intensification even easier to do. For example, if you give the insurance agent 10% more compensation if they sell more in the next 2 hours, they will be motivated to push themselves for those 2 hours and then see how many sales are possible. Then without the extra compensation, they can see how displaying this behavior would be good normally, as they will sell more, and make more.

Sometimes you may want to reduce behavior. You may want to reduce a behavior just once, and again, this is sometimes used as a step to a more permanent behavior change, in this case a step towards stopping the behavior all together. To reduce a behavior, you have to remove the trigger that prompts them to do it, and/or make the behavior harder to do (reduce the ability), and/or replace the motivator with de-motivators, for example punishment. However, you must be careful here, as de-motivators can lead to high dissatisfaction. You must weigh it up and see if it is not better to just make the action harder, and increase motivation of an alternative.

You may want them to commit to a behavior for a period of time. For them to do this, you will need to plan for the time period needed, you will need to boost motivation during this period, increase their ability and deliver triggers when both are high. You may need to have several triggers during the period of time built into your delivery plan. You should also frame the behavior in a way which reduces cost to them (time, effort or money) and increases benefits. Time the triggers when this is as optimal as possible. Fear here, can play a large role, as users can have concerns of what they do not know such as how long something will take, therefore you need to manage this by setting expectations so that the users mind doesn't run wild with thoughts! Another option here is to break it down into smaller behavioral changes or tasks that will increase their ability and be more achievable.

If the behavior is familiar, but you need it for a period of time, again it becomes easier as they already have knowledge of the behavior. However, you may need to ask more than once during the time period - they may need reminders and triggers. The challenge here is timing and ensuring that the triggers are not an annoyance.

Brand new behaviors are more often than not, never internalized on the first go. Getting people to do a new behavior for a period of time, as opposed to a complete behavior change is sometimes easier as there is a known end point. Using this as a step to behavior change, can be used to familiarize a person to a new behavior.

Now that you have some foundation in human behavior and points to consider when trying to get people to behave in a desired way, we can get more practical in how you could build your delivery plan.

BUILDING THE PLAN OF DELIVERY

There are three key pieces:

- Using data sources and completing a change impact analysis.
- Brand and marketing – when to use different vehicles and types such as transparency etc.
- Building the actual plan. Including vehicles, timing, user help etc.

Data Sources that Can Help

If you have been doing these user experience activities throughout you may have many sources of data available to you. If you haven't, don't worry, you will be able to obtain some from the designers or other members of the team.

For example:

Market Research: Perhaps a marketing team or HR team has demographics and other useful information for you to also take into consideration, e.g. to your population in a country that perhaps doesn't have great network coverage you may want to use a text email with no pictures or fancy formatting.

Concept Testing: This can you tell you a lot of information such as users' reactions to the ideas, what they are concerned about or like. You can then use this in the way that you communicate to them. For example, if you know a big concern of a website that deals with money transactions is security, you can ensure you focus on letting them know how secure the site is, hence dealing with the fear before it becomes a disabler of the behavior.

Usability Testing: This is useful to look at the reports and see which usability bugs were fixed and which were not. Sometimes during a project companies cannot fix everything as they make a decisions between criticality of a bug versus cost to fix, benefit of fixing and time. Using the usability testing reports and outcomes you can see what are the good features to perhaps point out, as well as what users found frustrating or confusing, and then ensure you give them direction or instruction to appease this. For example, if on a bank site users found it confusing that they had to enter their email for the user ID, and the company could not change the on screen text box from saying 'User ID', then in the sign up email, you could mention something like 'we have made it easy for you to remember your User ID - it is the email address you have registered for online banking'. Then, if you think back to the user flow, you should be thinking, what if they do not remember which email they used, and there was no way to check online at that same screen, you would also want to include here 'if you do not remember your email, do x'.

Change Impact Analysis

One of the key areas to understand when you are delivering something to end users and for MOC is understanding what is changing for them and how it will affect them. In many projects there is a change impact analysis, which is a simple methodology for defining this. If your project doesn't have one, then it is a good idea for you to do one as part of your delivery process. You will need to speak to other project members to gather these details such as the project manager or developers. There are many different forms of this analysis, some far more detailed than others, but I have found there are some key items that are important from an end user perspective that you should include, such as:

- AS-IS State: How is the end user's state now? In other words, what do they do today, as if your project isn't happening?
- TO-BE State: How will the end user be after they have your solution or your project launches.

- Segments impacted: Which segments are impacted? Specific groups that may have a change to their everyday activities.
- Actions changing for them: Per segment, which actions or tasks, processes or tools change for them, specifically?
- Gaps they may have: What knowledge do they not have today, that they may need for the TO-BE state?

Nothing explains it better than example, so figure 18 shows you a high level case for a new online process.

Once you have all the data to help you understand the changes that are coming to the different users, what could potentially be points of frustration or barriers to entry for them, the next layer is to understand how you will 'market' the solution to them. How will you persuade them to use it, ensure they can transition smoothly and are satisfied? The next layer is the branding and marketing of your solution during delivery.

AS-IS	TO-BE	TASK/ ACTIONS	ROLE/ SEGMENT	GAPS
Enter the data into spreadsheet.	Enter into the online Form.	Enter Data.	Admin	Train on process.
Calculate the monthly amounts..	Click "Calculate" button.	Run Report.		Training on Application - entry and calculation.
Create the graphical view.	Open the Application. Click on Views. Select View. Export to Presentation.	Present graph to management.	Sales Manager	Training on Application – create and export.

Table 19: Change impact analysis

BRAND & MARKETING

To know how best to market the solution to the users, it is smart to look at their current perception to know whether you have to reassure them, convince them, provoke thought or if it's an easy sell etc. to enable you to drive the adoption. Basically put - how will you win people over as much as possible the first time they hear about the solution. For example, if you are delivering a new sandwich service in an area where there is currently none, but you know many people in the area suggested one through a feedback mechanism - this would enable you to make a big 'splash' about the new sandwich service as it was something wanted by the audience, and you are delivering and answering their need.

You may already have some data to help, for example if you have completed perception testing (Chapter 6), or you knew some of what people thought from the concept testing. If you haven't, you may want to talk to a handful of people to get an idea of their current perception - very simply just asking them what they think about the solution, or if there is a current solution that is being replaced, getting a gauge on their current perception.

Marketing can be in many forms, you have the splashy style, the factual style, the awareness building, the prompting of thought, the building of anticipation, or the plan to build credibility, and more! You need to find out which style is best to drive the adoption of your solution. It may not just be one either, you may need a combination.

One thing is for certain, in my experience if you select the absolute wrong style many times, it becomes a joke just how much users will ignore the communications you give to them. For example, if you currently have a web page service that is always 'down' and they have a bad perception of it, then you create a plan about how wonderful the next release is, and that release hasn't really changed anything that was a pain point for them, I assure you - **they will see straight past your smoke and mirrors**. Actually what you may have needed here was transparency, to state very honestly that you acknowledge that availability of the service is not optimal, you are sorry, this is what you are doing about it, and in the meantime you have also fixed a, b and c. Many times, honesty is the best policy.

Here are some examples that will help you see the different types of delivery considerations. Sometimes you will use a mixture, or different kinds at different times.

	Situation	**Potential Plan**
Big cool bang	New coffee station – users have complained about this topic before.	Posters and Signs for launch. QR code to the new coffee type advert/information.
Create awareness	Improvement in email service downtime.	Newsletter - softer awareness campaign.

Create anticipation	A new product.	Mysterious ads, games involving it, draw for checking it out, posters, see and touch lunch.
Prompt thought	A food bank lunch to donate food.	Email, posters, video on charity.
Build trust	Café food improvements – was bad before.	Email admitting wrongs and stating the steps to fix. Table leaflets in café, digital displays.

Table 20: Delivery considerations

Building the Plan

With all of this information and thought, you actually have to build the plan. Now that you have all the information at your fingertips, have decided the core styles and messaging needed, you need to actually create the plan. This means thinking through the timing, and which vehicles are best to communicate the correct messages to the audience you have.

There are several examples of delivery plans out there, but it seems to me there are some core pieces of information that a basic one should have.

- Date - the date that this communication goes out, and for how long if it is physical e.g. a poster when it goes up and comes down, or if it's a commercial, the date and reoccurrence, frequency etc. If it is a Tweet, how frequent?
- Communication Purpose - what type is it? Is it creating awareness, is it giving information etc.?
- Key Messaging - what are the points to the communication?
- Audience - who is receiving it?
- Vehicle - how are you delivering it, e.g. an email, a poster, a video posted online, social media....?
- Ownership Over:
    - Creation - who is creating this communication?
    - Content - who must provide what content e.g. images, text, statistics etc.?
    - Approvals - is there anyone who must approve it? In companies there is usually some level of approval for anything going out to users - so who has to give the go ahead? It's always good to know how long this process takes and factor it into the plan.
    - Delivery - who owns the actual delivery e.g. sending the email, putting up the posters etc.?

It is important to realize just as the world changed many disciplines, communications and marketing were affected as well. Talent in this space should be just as up-to-date, especially concerning strategists. For example, they should understand how mobile plays a role, digital marketing and when to use which methods. They also need to understand their users. Often, those communicators that haven't kept up with the world do not always have the needed skill set to deal with experience as a whole. If you cannot hire for this, then think how you can help train existing resources to understand the world you are communicating in and the people you are communicating to. Personas and user flows can really help here, to ensure understanding of the user and tasks.

There are many things to consider here, and many things I have seen work, fail and heard from others in this business that I think are very important to know, especially when starting out on a user experience journey.

Fact 1 – Emails are everywhere... users are overwhelmed with emails and information from several departments, not just yours. They have email reminders, information and marketing from every function so they are very likely to ignore many of them. In fact, some people may even have emails like ones from HR or a particular email address for example, automatically removed to another folder. Not only do they have all these emails, they have all their actual work emails, and then if you think of a mobile device, they also have all their personal emails. Therefore one key learning I heard once was, sift through this over saturated environment and design it for silence. Use emails only when absolutely needed, think of other ways to get the messaging across, so that users know that if an email comes from you, they should at least have a glance.

Fact 2 - Communication is not synonymous to 'email'. If you look up the definition of communication, by the way some companies think, you would think 'email' was their dictionary definition of the word. Broaden your mind, to see the full communication landscape, what are all the potential vehicles your company has to offer you - list them all, and refer to it every time you are building a plan to pick the best one possible. For example, live meetings, posters, table leaflets, newsletters, meetings, lunch time talks, videos, digital media, social and the list continues....

Fact 3 – There is a perception of no time - that ever growing user perception that I have absolutely no time to deal with anything at all, especially not to read your email. Many times, I hear people say along the lines of, 'we only have about 40% of users adopting the tool'. I say 'what did you do to tell them to do it', they say 'oh, we emailed them, we also emailed them a reminder 2 weeks later'. The definition of insanity is said to be, repeating the same behavior over and over again and expecting a different outcome. Enough said.

Fact 4 - The design of the communication is as important as the words themselves. Sometimes, especially in less mature companies in the space, people do not think of communications as having a design element to them, but they do. Communication design is actually also very important for any type of written communication such as email or newsletter etc. You want to make sure your design is created for scanability. That is, a reader can look at it quickly and either take away the relevant information, or make the decision that they will spend more time reading it more thoroughly. On average you can expect, after actually making the decision to open the email or the link to a site etc., a reader spends just 5 seconds scanning, before making this decision. Just like design of the solution itself for the solution to be usable is important, design here is vital to manipulate the path of the readers gaze, to notice exactly what you want them to. The use of color, bolding, color blocks and images should be used carefully to enable this, while not making it look like a rainbow threw up on the page. You can even do a 'flash test' to see what information a user remembers or takes away with them after seeing your communication for 5 seconds – what stands out to them?

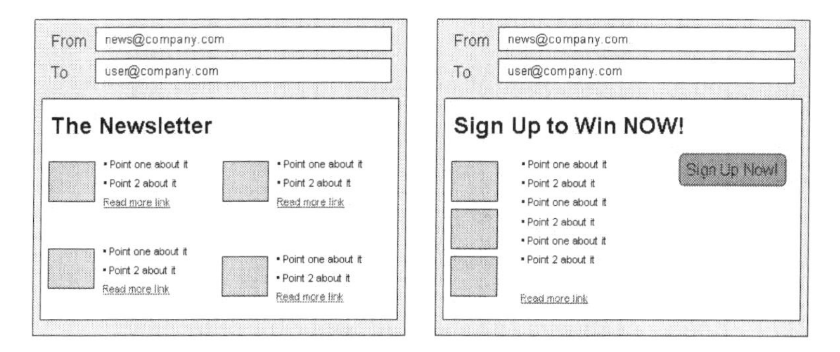

Figure 28 : Design mocks can be used in testing

Fact 5 - **Emails are a window of time**. They get sent at a particular moment, and that's that. For example, if you want to tell someone about a new service and you use an incentive to draw them to it such as the first 100 visitors to the site get entered into a draw, but the email recipient is on vacation and misses the date, then what? That's a non-disruptive example maybe, but think if you are changing a very important service, or for example users are enrolling in company stock. What if that email is missed? How else and where else will users get the same vital information if it is in fact very important and time sensitive.

Fact 6 – The correct method of communication is needed for the correct people. Next to your list of all the communications vehicles you have, you should start building data on who responds best to which types. For example, if you are launching something to the younger user base, perhaps a career development site aimed at people wanting a mentor - then you may choose to use a social media platform because they are readily checking their profiles and news on the company feed. If there is an outage that affects a core system, perhaps SMS is the best method for a group within the company that always have their mobile phone with them. You need to start collating the data on how best different people and groups respond to different communications methods to know which to use for the audience you are delivering to.

As always, these facts are great to know, this way you do not have to learn them by making mistakes first, however, an example is even better so table 20 demonstrates a high level delivery example so that you can see how to pull it all together. Be sure, to also think through the support for the user, how they will access it, any training for agents etc. You need to make sure that the end-to-end experience is thought through. You can use the user flows, to know when a user may need support or help, as well to identify any help information needed.

Date	Purpose	Key Message	Audience	Vehicle	Ownership		
					Create	Content	Approve
T- 6 weeks	Getting Support Ready	Knowledge for help desk agents and training. Articles for help ready.	Help Agents and Support Team	Online site, web training.	Support Team, Comms Team	Service Team	Service team, Support Manager
T – 3 weeks	Awareness	It is coming.	All	Communications Email	Comms Team	Service Team	Service Team, Executive
T – 2 weeks	Informative	Main points on this. Management support.	Managers	Video Exec Video Email	Executive Commination's Manager	Service Team	Executive
T – 1 weeks	Prompt Action	Training.	All	Brown Bag lunch Video Online	Comms Team, Service Team	Service Team	Service Team
Launch!	It's Here!	Announce launch. Reminder on training.	All		Comms Team	Service Team	Service Team
T + 1 week	Check in and survey	How the solution is going. Reminder on how to's and Training. Prompt Adoption.	All	Email	Comms Team	Service Team	Service Team
Monthly	Survey and Newsletter	Give news on improvements happened, adoption, whats coming.	Al	Email Newsletter	Comms Team	Service Team	Service Team

Table 21: Delivery plan

Now, you will need to flesh out your ongoing plan if one is needed, and as previously mentioned, ensure you have your mechanisms in place for ongoing feedback. You can refer back to when you decided which variables you would measure at the beginning to ensure these are also in the plan.

Your ongoing plan may be needed forever, for the life of the solution, or until a certain goal is achieved for example - you need people to register their contact details in the new HR system that is launched.

You will need a plan for the launch to get as many as possible, and then an ongoing plan until 100% of users have completed this task.

You will therefore need to consider two things. What data you will need to review at what frequency ongoing, and what you will do about it. With the HR system above, perhaps you will review monthly the following:

- Number of people that have completed the task.
- For those that have not, their demographical information - perhaps there is a pattern in location, team, department etc.

Then, on review of this, you will have to think through what you can do to drive the behavior you need.

Again, it is helpful to look through this with an example. It has been 1 month since you launched the HR system above. You receive the data on who has gone in and made the updates as shown in figure 28.

You then decide to do the following:

- Send a direct communications to Germany based people.
- Prompt Direct Managers of sales people to encourage their teams to adopt.

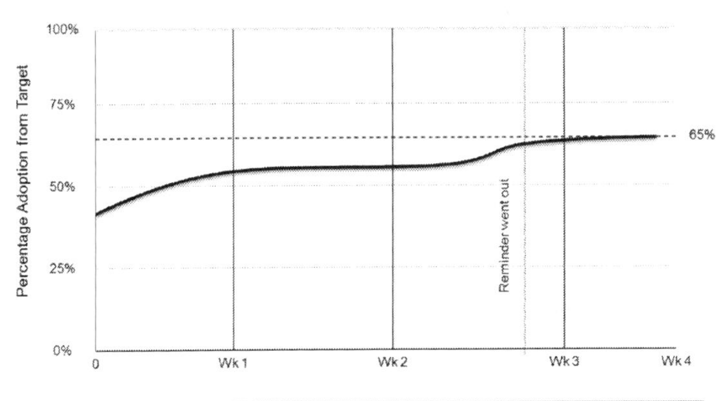

	Adoption Notes
Week 1	55% - All of North America Signed up
Week 2	55% - no change, no new sign ups
Week 3	63% - direct communications went to South America
Week 4	65% - none of Germany, low sales people adoption

Figure 29: Delivery data

After month 2, the data is again reviewed and it is found that there was an increase. Adoption from Sales was higher so clearly the action worked for that segment, however, Germany was still looking low on adoption.

It is worth a mention the impact of direct managers on enterprise user adoption. Executives are great to give the 'authorization and support', but direct managers supporting the new behavior or solution works wonders. Even better, is the manager using the solution themselves or with their team. In your delivery plan, it is good to plan around how the direct managers can help. For example, in the release of a new cloud solution to store documents, perhaps the manager will send weekly meeting notes to their employees via the tool – this demonstrates support, encouragement to use the tool and creates a constant action in which the employees will need to use the tool. This is similar to consumer solutions ensuring they are 'in the consumer's life' so to speak, and using figures of influence to endorse and use the solution.

From this data you see that the company-wide direct communication to Germany based people from month 1 didn't really work for that group so you decide to prompt local leaders to encourage their teams and offer an incentive.

You want to have a plan that everyone knows what data they are providing, to whom and who is driving the actions.

Data	**Owner** **Captures the data.**	**Send To** **Person who receives the data.**	**Owner** **Person who reviews the data and plans next steps.**	**Actions** **The actions to be taken.**
Users and amount	*NAME*	*NAME*	*NAME*	*ACTIONS*
Top Issues	*NAME*	*NAME*	*NAME*	*ACTIONS*

Table 22: Ongoing plan example – who is doing what and providing data

You see the pattern here, the key take away is this - you need to think what you do when you 'launch' the solution, and if needed, what your plans are ongoing to continue to drive the necessary behavior when an ongoing plan is needed. Types of data you may want to review on an ongoing basis include, number of users, sales, win/loss evaluation, support issues, number of support calls and reasons, user feedback and online identity (likes, recommendations, reviews, followers…).

Final Words

So there you have it, the framework and theory to get started with UX. I hope you have learned how UX plays an end-to-end role and are ready to go! I also hope that at least a piece of this has made you think of some facet of your own product, job or experience that you are delivering.

This was really meant as a UX primer for you to get started, get enough practical information and start your journey. There are so many good resources on every aspect from design to measures, so please spend some time reading more.

Simply put, the absolute keys to remember are to **know your users, their task and the flow** through the solution, and design for **beauty and function**. This works for almost anything. I'll leave you with a few examples I encountered recently to ponder the power of that triplet and how it plays a part in everything from planning, to how you validly test experience...

Example one is thinking through launching a new service at a tradeshow or conference.

The *Users* are conference-goers (tech savvy, unlikely to have their laptop with them, more likely to be carrying mobile devices).

The *Task* for the user is trying to learn from stands and booths that spark their interests.

The *Flow* is that the users will likely get drawn to a stand, spend less than 3min, unless something catches their attention, walk away to the next, look up the information later during a break – if they are interested.

This understanding leads you to 1) ensure they are drawn to your stand (thoughtful giveaways etc.) 2) ensure your giveaway leads to a mobile site that is well designed, that will get them to the next step towards buying.

Example two comes from a conversation I was having with a headset designer about the importance of the inline controls on a USB headset e.g. mute, power, volume given he was seeing users accidentally end calls when they mean to mute. Anyone in a corporation heavy in meetings knows that mute is essential! If you think of the user flow here the user has normally used another device to start and end the call e.g. client software on laptop or mobile device. The user is likely using a headset to have their hands-free, and usually are looking elsewhere e.g. a screen or on the road etc. Therefore they need to use the controls and know which is which without looking, so you would need to differentiate the buttons so that a user knows by touch. The key is to balance the sleek appeal and design with function. How can you test this specific feature? Here, blind tests could be used i.e. without sight, can a user mute a call, change the volume etc.?

Again, understand your user, their task and their flow, marry this with pleasing designs... and you are well on your way to creating beautiful, functional experiences.

Welcome to the UX world!

NUGGETS SUMMARY

Some key nuggets were bolded throughout the book. Here is a summary of them for your reference.

****

Experiences today have to be intuitively intelligent, adapting to the users situation, making them smarter, faster, more engaged and more efficient.

It has now become increasingly more evident that there is a strong relationship between a positive employee experience and a positive consumer experience.

When getting started, it is key to start out lean, prove the process works by demonstrating tangible value, and then start expanding throughout the organization.

The brand can only be born and cultivated through its value and through its demonstration of that value.

The key in today's world for experience is design for the goal and purpose, where usability is ensuring the user can do their task.

Experience is affected by everything the user interacts with. Perception is THEIR reality.

Know the needs versus listening to wants.

After one want is met, there is usually another want.

Users should see some value or return very quickly.

Don't lie to yourself as to where the creation of a great experience lies within the goals.

Why do something wrong and fix it later, if you can make it as good as it can be now?

Make the user faster and smarter in the completion of the task at hand.

How can you use known context to enable a better, adaptive experience depending on who the user is, their environment and what they are trying to do?

We are ultimately clouded by our own subset of experiences.

Fail fast and often, cheaply.

Users generally do not seem to like change unless the perceived value is high enough for them to switch or adopt a new behavior.

User experience research looks deeply into the entire user environment, who the users are, their needs, culture, and how everything interacts.

Each time, plan the best way to deliver the solution to the intended audience.

Human behavior is much more predictable than is generally thought, and their behavior is affected by the behavior of those in their networks.

The behavior of most people is likely to be regular and predictable.

Know your users, their task and their flow.

FRAMEWORK REFERENCE

Remember, which parts of the framework you implement, and to what extent, will depend on what you are trying to achieve with experience design. Here are the main parts of the framework for reference.

Roadmap and Strategy

- ☐ Experience vision.
- ☐ Data analysis (feedback, metrics, benchmark, segmentation, user needs…).
- ☐ Personas.
- ☐ Goals, constraints, requirements (Kano Model).
- ☐ Experience variables and baseline.
- ☐ Situations – who, what, why, where, when.
- ☐ Concept testing – if there are various options for solutions.
- ☐ Exemplar experience definition.
- ☐ Current solution usability testing or observations.

Ongoing

- ☐ What data is to be captured, by who and how?

- ☐ Who is responsible for actions as a result of this?
- ☐ With what frequency will the data be reviewed?

Actual

- ☐ User flows – you can start these as early as possible.
- ☐ Actual design – concepts may be considered.
- ☐ Concept testing – testing different ways to implement the solution.
- ☐ Wireframes, mocks or prototypes.
- ☐ Usability testing – can be iterative. Ensure each piece is usable, and the whole solution end-to-end is.

Delivery

- ☐ Change impact analysis – AS-IS, TO-BE states as well as gaps and plans to get there.
- ☐ Data – segments, personas, what we know of them.
- ☐ Communications and marketing plans – use the data you have, as well concept testing results and usability results.
- ☐ Support experience and considerations for any support readiness, SLA's, knowledge, how a user access's support etc.

REFERENCES AND READING

The Design Experience

Press and Cooper, 2003

The Design of Business: Why Design Thinking is the Next Competitive Advantage

Roger L. Martin, 2009

User Experience Research and Market Research, Human Factors International

Apala Lahiri Chavan, 2012

Cognition and the Intrinsic User Experience, UX Magazine

Jordan Julien, 2012

The Impact of the Internet on Human Behavior

Kim Krause Berg, 2009

Design of Emotional Experiences as a Source for Strategic Competitive Advantage

Lisa Cox, 2011

Design Research: Applied Exploration of People, Culture, Context and Form

Brenda Laurel, 2004

On the Collective Nature of Human Intelligence

Alexy Pentland, 2007

A Behavior Model for Persuasive Design

BJ Fogg, 2009

Human Behavior Representation Definition

Dr. Uwe Dompke, 2002

Reducing Variation In Human Behavior

Robert V Berry, Picard 1997

Ariety and Lowenstien, 2005

The Psychology Of Everyday Things

Donald Norman, 1988

***

http://www.nathan.com/thoughts/newmethods/scenarios/5.html

http://slideshare.net/vijay_bijaj/presentation-berry-premier-health

http://hd.media.mit.edu/tech_reports/tr-600.pdf

http://dtic.mil/cgi-bin/gettrdoc?ad=ada485573

http://searchengineland.com/the-impact-of-the-internet-on-human-behavior-209201

http://www.behaviorwizard.org/wp/all-previews-list/greenpath-behaviors-preview/

http://www.bjfogg.com/fbm_files/page4_1.pdf

http://www.refracted.co.uk/stephthesis.pdf

http://uxmag.com/articles/cognition-the-intrinsic-user-experience

http://blogs.forrester.com/kerry_bodine/12-11-02-designing_the_employee_experience

http://www.usability.gov/

http://www.agile-ux.com/2009/05/06/the-kano model%E2%80%A6-so-good-for-user-experience/

http://people.ucalgary.ca/~design/engg251/First%20Year%20 Files/kano.pdf

Made in the USA
Lexington, KY
07 February 2015